This book belongs to

_____
- -
_____

Grade: _____

School: _____

"A few minutes of math practice every day can help children master math skills. This '100+ Days of Timed Tests: Single & Double Digit Multiplication' is a beginner-level math practice workbook for kids in Grades 2 to 4. The workbook contains both 1-digit and 2-digit multiplication problems, designed for numbers up to 99. Kids should start with the 1-digit problems and progress to the 2-digit problems to build a strong foundation.

This set of math practice worksheets is designed to test multiplication skills both with and without regrouping. Kids can challenge themselves with the timed test problems, with the book mainly focusing on improving multiplication skills and building confidence levels. The book also contains answer key sheets at the end, making it easy to check kids' answers.

The book includes 25 problems to be solved daily, with a total of 101 pages of timed test practice sheets. It helps kids perform consistently and become excellent in multiplication. Please also check out our other higher-level math workbooks."

Example Problems:

Multiplication Without Regrouping

$$
\begin{array}{r}
31 \\
\times\ 23 \\
\hline
93 \\
+\ 62 \\
\hline
713
\end{array}
$$

31
23   ↑ 3x1=3; 3x3=9
93
31
23   ↑ 2x1=2; 2x3=6

62

Multiplication With Regrouping

$$
\begin{array}{r}
1\ \ 1 \\
45 \\
\times\ 23 \\
\hline
135 \\
+\ 90 \\
\hline
1035
\end{array}
$$

45
23   ↑ 3x5=15; 3x4=12 and add 1 => 13
135
45
23   ↑ 2x5=10; 2x4=8 and add 1 => 9

90

100+ Days of Timed Tests MULTIPLICATION SINGLE & DOUBLE DIGIT

2500+ Practice Problems

Practice Workbook

Ages 6 - 9

100+ Days

Grade 2-4

Math Drills

WITH & WITHOUT REGROUPING

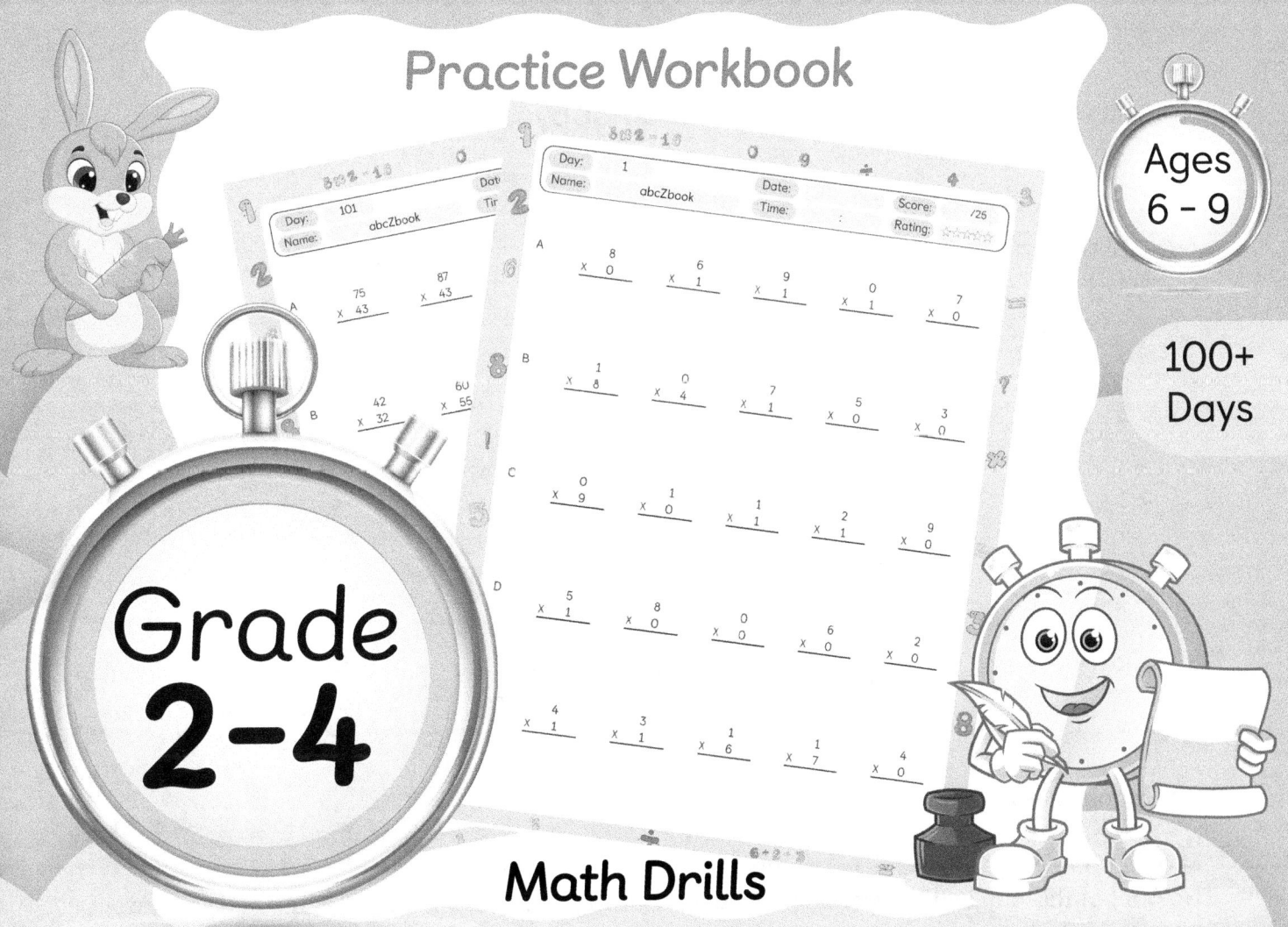

Greetings, wonderful parents!

Thank you for choosing this amazing book to help your child learn **Single & Double Digit Multiplication**. We hope you're just as excited as we are to begin this learning journey!

Your feedback is incredibly valuable to us. Please take a moment to leave a review on the platform where you purchased the book, and let us know what you thought. We're always striving to make our resources better and more effective, and your insights will help us do just that.

And, if you're ready for more learning adventures, check out our other books in the series. We promise they're just as fantastic as this one!

Thanks again for your support, and happy learning!

Best regards,
abcZbook Press
www.abczbook.com

abcZbook Press

Table of Contents

A

8	6	9	0	7
X 0	X 1	X 1	X 1	X 0

B

1	0	7	5	3
X 8	X 4	X 1	X 0	X 0

C

0	1	1	2	9
X 9	X 0	X 1	X 1	X 0

D

5	8	0	6	2
X 1	X 0	X 0	X 0	X 0

E

4	3	1	1	4
X 1	X 1	X 6	X 7	X 0

Day: 2

Date:

Score: /25

Name:

Time: :

Rating: ☆☆☆☆☆

A
$$\begin{array}{r} 5 \\ \times\ 3 \\ \hline \end{array}$$
$$\begin{array}{r} 2 \\ \times\ 3 \\ \hline \end{array}$$
$$\begin{array}{r} 1 \\ \times\ 3 \\ \hline \end{array}$$
$$\begin{array}{r} 0 \\ \times\ 3 \\ \hline \end{array}$$
$$\begin{array}{r} 5 \\ \times\ 2 \\ \hline \end{array}$$

B
$$\begin{array}{r} 2 \\ \times\ 5 \\ \hline \end{array}$$
$$\begin{array}{r} 3 \\ \times\ 0 \\ \hline \end{array}$$
$$\begin{array}{r} 9 \\ \times\ 3 \\ \hline \end{array}$$
$$\begin{array}{r} 8 \\ \times\ 2 \\ \hline \end{array}$$
$$\begin{array}{r} 4 \\ \times\ 2 \\ \hline \end{array}$$

C
$$\begin{array}{r} 3 \\ \times\ 8 \\ \hline \end{array}$$
$$\begin{array}{r} 7 \\ \times\ 2 \\ \hline \end{array}$$
$$\begin{array}{r} 7 \\ \times\ 3 \\ \hline \end{array}$$
$$\begin{array}{r} 4 \\ \times\ 3 \\ \hline \end{array}$$
$$\begin{array}{r} 6 \\ \times\ 2 \\ \hline \end{array}$$

D
$$\begin{array}{r} 3 \\ \times\ 3 \\ \hline \end{array}$$
$$\begin{array}{r} 2 \\ \times\ 2 \\ \hline \end{array}$$
$$\begin{array}{r} 0 \\ \times\ 2 \\ \hline \end{array}$$
$$\begin{array}{r} 9 \\ \times\ 2 \\ \hline \end{array}$$
$$\begin{array}{r} 3 \\ \times\ 2 \\ \hline \end{array}$$

E
$$\begin{array}{r} 6 \\ \times\ 3 \\ \hline \end{array}$$
$$\begin{array}{r} 8 \\ \times\ 3 \\ \hline \end{array}$$
$$\begin{array}{r} 2 \\ \times\ 6 \\ \hline \end{array}$$
$$\begin{array}{r} 2 \\ \times\ 4 \\ \hline \end{array}$$
$$\begin{array}{r} 1 \\ \times\ 2 \\ \hline \end{array}$$

A
$$\begin{array}{r}5\\ \times\ 5\\ \hline\end{array}\qquad\begin{array}{r}1\\ \times\ 5\\ \hline\end{array}\qquad\begin{array}{r}3\\ \times\ 5\\ \hline\end{array}\qquad\begin{array}{r}2\\ \times\ 5\\ \hline\end{array}\qquad\begin{array}{r}1\\ \times\ 4\\ \hline\end{array}$$

B
$$\begin{array}{r}5\\ \times\ 3\\ \hline\end{array}\qquad\begin{array}{r}4\\ \times\ 6\\ \hline\end{array}\qquad\begin{array}{r}7\\ \times\ 5\\ \hline\end{array}\qquad\begin{array}{r}0\\ \times\ 4\\ \hline\end{array}\qquad\begin{array}{r}5\\ \times\ 4\\ \hline\end{array}$$

C
$$\begin{array}{r}4\\ \times\ 5\\ \hline\end{array}\qquad\begin{array}{r}2\\ \times\ 4\\ \hline\end{array}\qquad\begin{array}{r}0\\ \times\ 5\\ \hline\end{array}\qquad\begin{array}{r}4\\ \times\ 5\\ \hline\end{array}\qquad\begin{array}{r}3\\ \times\ 4\\ \hline\end{array}$$

D
$$\begin{array}{r}8\\ \times\ 5\\ \hline\end{array}\qquad\begin{array}{r}9\\ \times\ 4\\ \hline\end{array}\qquad\begin{array}{r}7\\ \times\ 4\\ \hline\end{array}\qquad\begin{array}{r}4\\ \times\ 4\\ \hline\end{array}\qquad\begin{array}{r}6\\ \times\ 4\\ \hline\end{array}$$

E
$$\begin{array}{r}6\\ \times\ 5\\ \hline\end{array}\qquad\begin{array}{r}9\\ \times\ 5\\ \hline\end{array}\qquad\begin{array}{r}4\\ \times\ 7\\ \hline\end{array}\qquad\begin{array}{r}5\\ \times\ 2\\ \hline\end{array}\qquad\begin{array}{r}8\\ \times\ 4\\ \hline\end{array}$$

A
```
      7            6            2            0            3
  X   7        X   7        X   7        X   7        X   6
  _____      _____      _____      _____      _____
```

B
```
      7            6            3            9            7
  X   5        X   8        X   7        X   6        X   6
  _____      _____      _____      _____      _____
```

C
```
      6            5            4            9            2
  X   3        X   6        X   7        X   7        X   6
  _____      _____      _____      _____      _____
```

D
```
      5            6            4            0            1
  X   7        X   6        X   6        X   6        X   6
  _____      _____      _____      _____      _____
```

E
```
      1            8            6            7            8
  X   7        X   7        X   7        X   4        X   6
  _____      _____      _____      _____      _____
```

A

9	7	0	3	5
× 9	× 9	× 9	× 9	× 8

B

9	8	2	9	0
× 4	× 0	× 9	× 8	× 8

C

8	1	1	4	7
× 1	× 8	× 9	× 9	× 8

D

6	8	6	3	2
× 9	× 8	× 8	× 8	× 8

E

5	8	8	9	4
× 9	× 9	× 8	× 6	× 8

A
$$\begin{array}{r} 2 \\ \times\ 1 \\ \hline \end{array}$$
$$\begin{array}{r} 5 \\ \times\ 1 \\ \hline \end{array}$$
$$\begin{array}{r} 3 \\ \times\ 1 \\ \hline \end{array}$$
$$\begin{array}{r} 8 \\ \times\ 1 \\ \hline \end{array}$$
$$\begin{array}{r} 9 \\ \times\ 0 \\ \hline \end{array}$$

B
$$\begin{array}{r} 1 \\ \times\ 5 \\ \hline \end{array}$$
$$\begin{array}{r} 0 \\ \times\ 2 \\ \hline \end{array}$$
$$\begin{array}{r} 9 \\ \times\ 1 \\ \hline \end{array}$$
$$\begin{array}{r} 4 \\ \times\ 1 \\ \hline \end{array}$$
$$\begin{array}{r} 7 \\ \times\ 0 \\ \hline \end{array}$$

C
$$\begin{array}{r} 0 \\ \times\ 4 \\ \hline \end{array}$$
$$\begin{array}{r} 8 \\ \times\ 0 \\ \hline \end{array}$$
$$\begin{array}{r} 1 \\ \times\ 1 \\ \hline \end{array}$$
$$\begin{array}{r} 1 \\ \times\ 3 \\ \hline \end{array}$$
$$\begin{array}{r} 6 \\ \times\ 0 \\ \hline \end{array}$$

D
$$\begin{array}{r} 1 \\ \times\ 7 \\ \hline \end{array}$$
$$\begin{array}{r} 4 \\ \times\ 0 \\ \hline \end{array}$$
$$\begin{array}{r} 2 \\ \times\ 0 \\ \hline \end{array}$$
$$\begin{array}{r} 1 \\ \times\ 0 \\ \hline \end{array}$$
$$\begin{array}{r} 5 \\ \times\ 0 \\ \hline \end{array}$$

E
$$\begin{array}{r} 6 \\ \times\ 1 \\ \hline \end{array}$$
$$\begin{array}{r} 7 \\ \times\ 1 \\ \hline \end{array}$$
$$\begin{array}{r} 1 \\ \times\ 6 \\ \hline \end{array}$$
$$\begin{array}{r} 1 \\ \times\ 1 \\ \hline \end{array}$$
$$\begin{array}{r} 3 \\ \times\ 0 \\ \hline \end{array}$$

A
$$\begin{array}{r} 8 \\ \times\ 3 \\ \hline \end{array}$$
$$\begin{array}{r} 1 \\ \times\ 3 \\ \hline \end{array}$$
$$\begin{array}{r} 2 \\ \times\ 3 \\ \hline \end{array}$$
$$\begin{array}{r} 5 \\ \times\ 3 \\ \hline \end{array}$$
$$\begin{array}{r} 4 \\ \times\ 2 \\ \hline \end{array}$$

B
$$\begin{array}{r} 2 \\ \times\ 5 \\ \hline \end{array}$$
$$\begin{array}{r} 3 \\ \times\ 6 \\ \hline \end{array}$$
$$\begin{array}{r} 3 \\ \times\ 3 \\ \hline \end{array}$$
$$\begin{array}{r} 4 \\ \times\ 2 \\ \hline \end{array}$$
$$\begin{array}{r} 8 \\ \times\ 2 \\ \hline \end{array}$$

C
$$\begin{array}{r} 3 \\ \times\ 7 \\ \hline \end{array}$$
$$\begin{array}{r} 1 \\ \times\ 2 \\ \hline \end{array}$$
$$\begin{array}{r} 9 \\ \times\ 3 \\ \hline \end{array}$$
$$\begin{array}{r} 3 \\ \times\ 9 \\ \hline \end{array}$$
$$\begin{array}{r} 3 \\ \times\ 2 \\ \hline \end{array}$$

D
$$\begin{array}{r} 3 \\ \times\ 8 \\ \hline \end{array}$$
$$\begin{array}{r} 9 \\ \times\ 2 \\ \hline \end{array}$$
$$\begin{array}{r} 5 \\ \times\ 2 \\ \hline \end{array}$$
$$\begin{array}{r} 6 \\ \times\ 2 \\ \hline \end{array}$$
$$\begin{array}{r} 2 \\ \times\ 2 \\ \hline \end{array}$$

E
$$\begin{array}{r} 7 \\ \times\ 3 \\ \hline \end{array}$$
$$\begin{array}{r} 6 \\ \times\ 3 \\ \hline \end{array}$$
$$\begin{array}{r} 2 \\ \times\ 4 \\ \hline \end{array}$$
$$\begin{array}{r} 2 \\ \times\ 3 \\ \hline \end{array}$$
$$\begin{array}{r} 7 \\ \times\ 2 \\ \hline \end{array}$$

A

```
    6          3          4          9          9
X   5      X   5      X   5      X   5      X   4
```

B

```
    5          4          8          2          1
X   8      X   5      X   5      X   4      X   4
```

C

```
    4          4          1          5          2
X   3      X   4      X   5      X   4      X   4
```

D

```
    5          8          6          3          7
X   7      X   4      X   4      X   4      X   4
```

E

```
    7          5          4          5          5
X   5      X   5      X   9      X   6      X   4
```

A

$$\begin{array}{r} 8 \\ \times\ 7 \\ \hline \end{array}$$
$$\begin{array}{r} 2 \\ \times\ 7 \\ \hline \end{array}$$
$$\begin{array}{r} 7 \\ \times\ 7 \\ \hline \end{array}$$
$$\begin{array}{r} 6 \\ \times\ 7 \\ \hline \end{array}$$
$$\begin{array}{r} 1 \\ \times\ 6 \\ \hline \end{array}$$

B

$$\begin{array}{r} 7 \\ \times\ 8 \\ \hline \end{array}$$
$$\begin{array}{r} 6 \\ \times\ 5 \\ \hline \end{array}$$
$$\begin{array}{r} 9 \\ \times\ 7 \\ \hline \end{array}$$
$$\begin{array}{r} 1 \\ \times\ 6 \\ \hline \end{array}$$
$$\begin{array}{r} 5 \\ \times\ 6 \\ \hline \end{array}$$

C

$$\begin{array}{r} 6 \\ \times\ 1 \\ \hline \end{array}$$
$$\begin{array}{r} 8 \\ \times\ 6 \\ \hline \end{array}$$
$$\begin{array}{r} 3 \\ \times\ 7 \\ \hline \end{array}$$
$$\begin{array}{r} 7 \\ \times\ 4 \\ \hline \end{array}$$
$$\begin{array}{r} 4 \\ \times\ 6 \\ \hline \end{array}$$

D

$$\begin{array}{r} 7 \\ \times\ 6 \\ \hline \end{array}$$
$$\begin{array}{r} 6 \\ \times\ 6 \\ \hline \end{array}$$
$$\begin{array}{r} 7 \\ \times\ 6 \\ \hline \end{array}$$
$$\begin{array}{r} 2 \\ \times\ 6 \\ \hline \end{array}$$
$$\begin{array}{r} 3 \\ \times\ 6 \\ \hline \end{array}$$

E

$$\begin{array}{r} 5 \\ \times\ 7 \\ \hline \end{array}$$
$$\begin{array}{r} 4 \\ \times\ 7 \\ \hline \end{array}$$
$$\begin{array}{r} 6 \\ \times\ 9 \\ \hline \end{array}$$
$$\begin{array}{r} 7 \\ \times\ 7 \\ \hline \end{array}$$
$$\begin{array}{r} 9 \\ \times\ 6 \\ \hline \end{array}$$

A
$$
\begin{array}{r} 9 \\ \times\ 9 \\ \hline \end{array}
\quad
\begin{array}{r} 8 \\ \times\ 9 \\ \hline \end{array}
\quad
\begin{array}{r} 3 \\ \times\ 9 \\ \hline \end{array}
\quad
\begin{array}{r} 4 \\ \times\ 9 \\ \hline \end{array}
\quad
\begin{array}{r} 7 \\ \times\ 8 \\ \hline \end{array}
$$

B
$$
\begin{array}{r} 9 \\ \times\ 1 \\ \hline \end{array}
\quad
\begin{array}{r} 8 \\ \times\ 9 \\ \hline \end{array}
\quad
\begin{array}{r} 5 \\ \times\ 9 \\ \hline \end{array}
\quad
\begin{array}{r} 1 \\ \times\ 8 \\ \hline \end{array}
\quad
\begin{array}{r} 9 \\ \times\ 8 \\ \hline \end{array}
$$

C
$$
\begin{array}{r} 8 \\ \times\ 2 \\ \hline \end{array}
\quad
\begin{array}{r} 1 \\ \times\ 8 \\ \hline \end{array}
\quad
\begin{array}{r} 7 \\ \times\ 9 \\ \hline \end{array}
\quad
\begin{array}{r} 9 \\ \times\ 7 \\ \hline \end{array}
\quad
\begin{array}{r} 2 \\ \times\ 8 \\ \hline \end{array}
$$

D
$$
\begin{array}{r} 9 \\ \times\ 8 \\ \hline \end{array}
\quad
\begin{array}{r} 6 \\ \times\ 8 \\ \hline \end{array}
\quad
\begin{array}{r} 8 \\ \times\ 8 \\ \hline \end{array}
\quad
\begin{array}{r} 3 \\ \times\ 8 \\ \hline \end{array}
\quad
\begin{array}{r} 4 \\ \times\ 8 \\ \hline \end{array}
$$

E
$$
\begin{array}{r} 2 \\ \times\ 9 \\ \hline \end{array}
\quad
\begin{array}{r} 6 \\ \times\ 9 \\ \hline \end{array}
\quad
\begin{array}{r} 8 \\ \times\ 4 \\ \hline \end{array}
\quad
\begin{array}{r} 9 \\ \times\ 5 \\ \hline \end{array}
\quad
\begin{array}{r} 5 \\ \times\ 8 \\ \hline \end{array}
$$

A
$$\begin{array}{r} 6 \\ \times\ 2 \\ \hline \end{array}$$
$$\begin{array}{r} 2 \\ \times\ 2 \\ \hline \end{array}$$
$$\begin{array}{r} 8 \\ \times\ 2 \\ \hline \end{array}$$
$$\begin{array}{r} 5 \\ \times\ 2 \\ \hline \end{array}$$
$$\begin{array}{r} 5 \\ \times\ 1 \\ \hline \end{array}$$

B
$$\begin{array}{r} 2 \\ \times\ 6 \\ \hline \end{array}$$
$$\begin{array}{r} 1 \\ \times\ 9 \\ \hline \end{array}$$
$$\begin{array}{r} 4 \\ \times\ 2 \\ \hline \end{array}$$
$$\begin{array}{r} 9 \\ \times\ 1 \\ \hline \end{array}$$
$$\begin{array}{r} 10 \\ \times\ 1 \\ \hline \end{array}$$

C
$$\begin{array}{r} 1 \\ \times\ 3 \\ \hline \end{array}$$
$$\begin{array}{r} 1 \\ \times\ 1 \\ \hline \end{array}$$
$$\begin{array}{r} 10 \\ \times\ 2 \\ \hline \end{array}$$
$$\begin{array}{r} 7 \\ \times\ 2 \\ \hline \end{array}$$
$$\begin{array}{r} 8 \\ \times\ 1 \\ \hline \end{array}$$

D
$$\begin{array}{r} 1 \\ \times\ 2 \\ \hline \end{array}$$
$$\begin{array}{r} 3 \\ \times\ 1 \\ \hline \end{array}$$
$$\begin{array}{r} 4 \\ \times\ 1 \\ \hline \end{array}$$
$$\begin{array}{r} 2 \\ \times\ 1 \\ \hline \end{array}$$
$$\begin{array}{r} 6 \\ \times\ 1 \\ \hline \end{array}$$

E
$$\begin{array}{r} 9 \\ \times\ 2 \\ \hline \end{array}$$
$$\begin{array}{r} 3 \\ \times\ 2 \\ \hline \end{array}$$
$$\begin{array}{r} 1 \\ \times\ 4 \\ \hline \end{array}$$
$$\begin{array}{r} 2 \\ \times\ 1 \\ \hline \end{array}$$
$$\begin{array}{r} 7 \\ \times\ 1 \\ \hline \end{array}$$

A

8	1	9	2	8
x 4	x 4	x 4	x 4	x 3

B

4	3	3	6	3
x 3	x 9	x 4	x 3	x 3

C

3	10	4	5	1
x 8	x 3	x 4	x 4	x 3

D

7	9	5	4	7
x 4	x 3	x 3	x 3	x 3

E

6	10	3	4	2
x 4	x 4	x 7	x 4	x 3

A

	4		7		3		8		1
X	6	X	6	X	6	X	6	X	5

B

	6		5		6		8		10
X	1	X	2	X	6	X	5	X	5

C

	5		6		5		10		2
X	4	X	5	X	6	X	6	X	5

D

	2		7		9		3		5
X	6	X	5	X	5	X	5	X	5

E

	9		1		5		6		4
X	6	X	6	X	8	X	6	X	5

A

$$10 \times 8$$ $$3 \times 8$$ $$1 \times 8$$ $$4 \times 8$$ $$7 \times 7$$

B

$$8 \times 9$$ $$7 \times 2$$ $$9 \times 8$$ $$3 \times 7$$ $$1 \times 7$$

C

$$7 \times 7$$ $$5 \times 7$$ $$2 \times 8$$ $$6 \times 8$$ $$4 \times 7$$

D

$$7 \times 8$$ $$8 \times 7$$ $$9 \times 7$$ $$6 \times 7$$ $$10 \times 7$$

E

$$5 \times 8$$ $$8 \times 8$$ $$7 \times 4$$ $$8 \times 5$$ $$2 \times 7$$

A
```
     10        10        10        10         3
x     6    x     4    x     8    x     3    x     9
```

B
```
     10         9        10         5        10
x     8    x     4    x     1    x     9    x     9
```

C
```
      9         6        10        10         9
x     2    x     9    x    10    x     5    x     9
```

D
```
     10         2         4         1         8
x     7    x     9    x     9    x     9    x     9
```

E
```
     10        10         9        10         7
x     9    x     2    x     5    x     6    x     9
```

A
```
    10          3          8          7          6
x    1      x   0      x   0      x   0      x   1
```

B
```
     4          1          2          9          5
x    1      x   1      x   1      x   0      x   1
```

C
```
     7          6          9          5          8
x    0      x   0      x   0      x   1      x   0
```

D
```
     1         10          2          3          4
x    1      x   0      x   1      x   1      x   0
```

E
```
     8          6          5          9         10
x    0      x   0      x   0      x   1      x   1
```

A
```
     8          9          4          1          6
x    1     x    2     x    1     x    1     x    1
_____   _____   _____   _____   _____
```

B
```
     7          3          5         10          2
x    2     x    2     x    2     x    2     x    1
_____   _____   _____   _____   _____
```

C
```
    10          2          8          7          9
x    1     x    2     x    2     x    2     x    1
_____   _____   _____   _____   _____
```

D
```
     4          5          6          3          1
x    1     x    1     x    2     x    1     x    2
_____   _____   _____   _____   _____
```

E
```
     2          8          9          3         10
x    1     x    1     x    1     x    2     x    2
_____   _____   _____   _____   _____
```

A

4	5	1	7	6
X 1	X 1	X 2	X 1	X 1

B

8	2	10	3	9
X 3	X 1	X 3	X 3	X 2

C

5	9	10	4	8
X 1	X 1	X 3	X 3	X 1

D

6	7	2	1	3
X 3	X 3	X 3	X 1	X 3

E

2	3	8	6	10
X 3	X 3	X 1	X 1	X 2

A

$$\begin{array}{r} 9 \\ \times\ 1 \\ \hline \end{array}$$
$$\begin{array}{r} 2 \\ \times\ 3 \\ \hline \end{array}$$
$$\begin{array}{r} 4 \\ \times\ 1 \\ \hline \end{array}$$
$$\begin{array}{r} 6 \\ \times\ 2 \\ \hline \end{array}$$
$$\begin{array}{r} 3 \\ \times\ 2 \\ \hline \end{array}$$

B

$$\begin{array}{r} 10 \\ \times\ 4 \\ \hline \end{array}$$
$$\begin{array}{r} 5 \\ \times\ 1 \\ \hline \end{array}$$
$$\begin{array}{r} 1 \\ \times\ 2 \\ \hline \end{array}$$
$$\begin{array}{r} 7 \\ \times\ 4 \\ \hline \end{array}$$
$$\begin{array}{r} 8 \\ \times\ 3 \\ \hline \end{array}$$

C

$$\begin{array}{r} 10 \\ \times\ 1 \\ \hline \end{array}$$
$$\begin{array}{r} 8 \\ \times\ 4 \\ \hline \end{array}$$
$$\begin{array}{r} 1 \\ \times\ 4 \\ \hline \end{array}$$
$$\begin{array}{r} 9 \\ \times\ 4 \\ \hline \end{array}$$
$$\begin{array}{r} 5 \\ \times\ 1 \\ \hline \end{array}$$

D

$$\begin{array}{r} 3 \\ \times\ 3 \\ \hline \end{array}$$
$$\begin{array}{r} 2 \\ \times\ 3 \\ \hline \end{array}$$
$$\begin{array}{r} 7 \\ \times\ 2 \\ \hline \end{array}$$
$$\begin{array}{r} 6 \\ \times\ 3 \\ \hline \end{array}$$
$$\begin{array}{r} 4 \\ \times\ 2 \\ \hline \end{array}$$

E

$$\begin{array}{r} 2 \\ \times\ 2 \\ \hline \end{array}$$
$$\begin{array}{r} 9 \\ \times\ 3 \\ \hline \end{array}$$
$$\begin{array}{r} 8 \\ \times\ 4 \\ \hline \end{array}$$
$$\begin{array}{r} 6 \\ \times\ 3 \\ \hline \end{array}$$
$$\begin{array}{r} 4 \\ \times\ 2 \\ \hline \end{array}$$

Day:	20	Date:		Score:	/25
Name:		Time:	:	Rating:	☆☆☆☆☆☆

A

$$\begin{array}{r} 8 \\ \times\ 3 \\ \hline \end{array} \qquad \begin{array}{r} 2 \\ \times\ 2 \\ \hline \end{array} \qquad \begin{array}{r} 7 \\ \times\ 1 \\ \hline \end{array} \qquad \begin{array}{r} 5 \\ \times\ 4 \\ \hline \end{array} \qquad \begin{array}{r} 10 \\ \times\ 2 \\ \hline \end{array}$$

B

$$\begin{array}{r} 1 \\ \times\ 5 \\ \hline \end{array} \qquad \begin{array}{r} 3 \\ \times\ 3 \\ \hline \end{array} \qquad \begin{array}{r} 6 \\ \times\ 5 \\ \hline \end{array} \qquad \begin{array}{r} 9 \\ \times\ 1 \\ \hline \end{array} \qquad \begin{array}{r} 4 \\ \times\ 1 \\ \hline \end{array}$$

C

$$\begin{array}{r} 2 \\ \times\ 5 \\ \hline \end{array} \qquad \begin{array}{r} 5 \\ \times\ 4 \\ \hline \end{array} \qquad \begin{array}{r} 3 \\ \times\ 5 \\ \hline \end{array} \qquad \begin{array}{r} 10 \\ \times\ 5 \\ \hline \end{array} \qquad \begin{array}{r} 4 \\ \times\ 2 \\ \hline \end{array}$$

D

$$\begin{array}{r} 8 \\ \times\ 3 \\ \hline \end{array} \qquad \begin{array}{r} 7 \\ \times\ 1 \\ \hline \end{array} \qquad \begin{array}{r} 6 \\ \times\ 2 \\ \hline \end{array} \qquad \begin{array}{r} 9 \\ \times\ 3 \\ \hline \end{array} \qquad \begin{array}{r} 1 \\ \times\ 4 \\ \hline \end{array}$$

E

$$\begin{array}{r} 2 \\ \times\ 3 \\ \hline \end{array} \qquad \begin{array}{r} 3 \\ \times\ 3 \\ \hline \end{array} \qquad \begin{array}{r} 8 \\ \times\ 5 \\ \hline \end{array} \qquad \begin{array}{r} 6 \\ \times\ 4 \\ \hline \end{array} \qquad \begin{array}{r} 10 \\ \times\ 1 \\ \hline \end{array}$$

A

7	6	5	4	8
X 5	X 5	X 1	X 2	X 2

B

2	10	1	3	9
X 6	X 1	X 1	X 5	X 5

C

9	5	3	4	1
X 1	X 2	X 5	X 6	X 2

D

10	6	7	8	2
X 5	X 2	X 6	X 1	X 1

E

4	10	9	2	8
X 1	X 5	X 5	X 1	X 4

A
$\begin{array}{r} 8 \\ \times\ 7 \\ \hline \end{array}$
$\begin{array}{r} 1 \\ \times\ 3 \\ \hline \end{array}$
$\begin{array}{r} 10 \\ \times\ 1 \\ \hline \end{array}$
$\begin{array}{r} 2 \\ \times\ 3 \\ \hline \end{array}$
$\begin{array}{r} 7 \\ \times\ 5 \\ \hline \end{array}$

B
$\begin{array}{r} 5 \\ \times\ 5 \\ \hline \end{array}$
$\begin{array}{r} 9 \\ \times\ 1 \\ \hline \end{array}$
$\begin{array}{r} 6 \\ \times\ 6 \\ \hline \end{array}$
$\begin{array}{r} 3 \\ \times\ 6 \\ \hline \end{array}$
$\begin{array}{r} 4 \\ \times\ 2 \\ \hline \end{array}$

C
$\begin{array}{r} 10 \\ \times\ 6 \\ \hline \end{array}$
$\begin{array}{r} 5 \\ \times\ 7 \\ \hline \end{array}$
$\begin{array}{r} 4 \\ \times\ 5 \\ \hline \end{array}$
$\begin{array}{r} 7 \\ \times\ 2 \\ \hline \end{array}$
$\begin{array}{r} 8 \\ \times\ 4 \\ \hline \end{array}$

D
$\begin{array}{r} 1 \\ \times\ 3 \\ \hline \end{array}$
$\begin{array}{r} 3 \\ \times\ 2 \\ \hline \end{array}$
$\begin{array}{r} 9 \\ \times\ 7 \\ \hline \end{array}$
$\begin{array}{r} 6 \\ \times\ 2 \\ \hline \end{array}$
$\begin{array}{r} 2 \\ \times\ 3 \\ \hline \end{array}$

E
$\begin{array}{r} 5 \\ \times\ 4 \\ \hline \end{array}$
$\begin{array}{r} 7 \\ \times\ 1 \\ \hline \end{array}$
$\begin{array}{r} 6 \\ \times\ 5 \\ \hline \end{array}$
$\begin{array}{r} 9 \\ \times\ 5 \\ \hline \end{array}$
$\begin{array}{r} 2 \\ \times\ 4 \\ \hline \end{array}$

A
```
     10          2          4          3          9
   X  7        X  7      X  4      X  8      X  5
```

B
```
      1          6          7          5          8
   X  4        X  6      X  6      X  5      X  6
```

C
```
      6          3          9          4         10
   X  2        X  1      X  3      X  5      X  2
```

D
```
      1          8          2          7          5
   X  8        X  7      X  4      X  3      X  3
```

E
```
      4          9          2          6          1
   X  8        X  1      X  3      X  4      X  6
```

A

7	5	6	9	10
× 9	× 5	× 2	× 4	× 6

B

8	3	1	2	4
× 8	× 6	× 3	× 6	× 2

C

10	3	1	9	5
× 5	× 4	× 7	× 7	× 8

D

2	6	8	7	4
× 4	× 5	× 3	× 2	× 4

E

1	2	8	7	4
× 6	× 7	× 5	× 6	× 7

A

4	7	10	5	9
X 7	X 5	X 8	X 3	X 4

B

1	3	10	2	6
X 9	X 6	X 2	X 9	X 1

C

9	1	3	2	10
X 2	X 5	X 4	X 3	X 6

D

7	8	5	10	4
X 6	X 7	X 8	X 1	X 3

E

6	10	2	10	3
X 2	X 7	X 2	X 4	X 8

A
```
    7          5         10          4          2
X   8      X   1      X   5      X   2      X   7
_____    _____    _____    _____    _____
```

B
```
    3          9          8          6          1
X   6      X   4      X   9      X   3      X  10
_____    _____    _____    _____    _____
```

C
```
    5          7          8          6         10
X   8      X   2      X   5      X   4      X   9
_____    _____    _____    _____    _____
```

D
```
    2          9          4          1          9
X   7      X   7      X   3      X   6      X   1
_____    _____    _____    _____    _____
```

E
```
    8         10          1         10          5
X   3      X   5      X   6      X  10      X   8
_____    _____    _____    _____    _____
```

A

3	10	8	1	7
× 6	× 2	× 5	× 4	× 1

B

5	2	9	6	4
× 3	× 9	× 7	× 8	× 4

C

7	3	10	5	1
× 4	× 2	× 8	× 2	× 1

D

2	8	4	9	6
× 5	× 3	× 6	× 9	× 7

E

8	4	7	5	3
× 8	× 5	× 2	× 9	× 1

Day: 28

Name:

Date:

Time: :

Score: /25

Rating: ☆☆☆☆☆☆

A
$$\begin{array}{r} 4 \\ \times\ 2 \\ \hline \end{array}$$
$$\begin{array}{r} 8 \\ \times\ 4 \\ \hline \end{array}$$
$$\begin{array}{r} 2 \\ \times\ 1 \\ \hline \end{array}$$
$$\begin{array}{r} 1 \\ \times\ 6 \\ \hline \end{array}$$
$$\begin{array}{r} 5 \\ \times\ 7 \\ \hline \end{array}$$

B
$$\begin{array}{r} 6 \\ \times\ 8 \\ \hline \end{array}$$
$$\begin{array}{r} 10 \\ \times\ 5 \\ \hline \end{array}$$
$$\begin{array}{r} 9 \\ \times\ 9 \\ \hline \end{array}$$
$$\begin{array}{r} 3 \\ \times\ 3 \\ \hline \end{array}$$
$$\begin{array}{r} 10 \\ \times\ 7 \\ \hline \end{array}$$

C
$$\begin{array}{r} 7 \\ \times\ 8 \\ \hline \end{array}$$
$$\begin{array}{r} 3 \\ \times\ 9 \\ \hline \end{array}$$
$$\begin{array}{r} 2 \\ \times\ 3 \\ \hline \end{array}$$
$$\begin{array}{r} 6 \\ \times\ 4 \\ \hline \end{array}$$
$$\begin{array}{r} 10 \\ \times\ 4 \\ \hline \end{array}$$

D
$$\begin{array}{r} 4 \\ \times\ 2 \\ \hline \end{array}$$
$$\begin{array}{r} 8 \\ \times\ 7 \\ \hline \end{array}$$
$$\begin{array}{r} 9 \\ \times\ 5 \\ \hline \end{array}$$
$$\begin{array}{r} 1 \\ \times\ 6 \\ \hline \end{array}$$
$$\begin{array}{r} 5 \\ \times\ 1 \\ \hline \end{array}$$

E
$$\begin{array}{r} 1 \\ \times\ 7 \\ \hline \end{array}$$
$$\begin{array}{r} 7 \\ \times\ 2 \\ \hline \end{array}$$
$$\begin{array}{r} 6 \\ \times\ 4 \\ \hline \end{array}$$
$$\begin{array}{r} 3 \\ \times\ 1 \\ \hline \end{array}$$
$$\begin{array}{r} 5 \\ \times\ 6 \\ \hline \end{array}$$

A

$$3 \times 3$$ $$5 \times 9$$ $$4 \times 2$$ $$8 \times 3$$ $$9 \times 7$$

B

$$2 \times 5$$ $$7 \times 4$$ $$6 \times 6$$ $$1 \times 8$$ $$10 \times 1$$

C

$$5 \times 5$$ $$3 \times 8$$ $$2 \times 2$$ $$4 \times 7$$ $$7 \times 1$$

D

$$8 \times 4$$ $$9 \times 9$$ $$6 \times 2$$ $$1 \times 3$$ $$10 \times 6$$

E

$$4 \times 4$$ $$6 \times 9$$ $$9 \times 3$$ $$10 \times 5$$ $$1 \times 6$$

A
```
     4          9          5          3          7
x    9     x    1     x    2     x    3     x    8
_____   _____   _____   _____   _____
```

B
```
     6          2          8         10          1
x    6     x    7     x    6     x    5     x    2
_____   _____   _____   _____   _____
```

C
```
     1          4          2          7          5
x    5     x    9     x    8     x    7     x    1
_____   _____   _____   _____   _____
```

D
```
     9          3          8          6         10
x    7     x    8     x    3     x    4     x    4
_____   _____   _____   _____   _____
```

E
```
    10          3          7          4          6
x    6     x    5     x    3     x    1     x    9
_____   _____   _____   _____   _____
```

A
```
     6          5          8         10          2
  X  5       X  1       X  4       X  7       X  8
```

B
```
     9          3          4          1         10
  X  6       X  9       X  3       X  2       X  7
```

C
```
     1          2          9          7          5
  X  6       X  4       X  5       X  2       X  5
```

D
```
     4          8         10          6          3
  X  8       X  9       X  1       X  7       X  3
```

E
```
    10          7          3          5          2
  X  8       X  7       X  3       X  4       X  3
```

A
```
    6         1         7        10         2
x   9     x   4     x   1     x   4     x   3
_____  _____  _____  _____  _____
```

B
```
    9         4         8         5         3
x   2     x   6     x   8     x   7     x   5
_____  _____  _____  _____  _____
```

C
```
    8        10         2         9         4
x   6     x   7     x   3     x   3     x   2
_____  _____  _____  _____  _____
```

D
```
    7         6         5         3         1
x   8     x   7     x   5     x   4     x   8
_____  _____  _____  _____  _____
```

E
```
    3         5         7         9         6
x   2     x   1     x   1     x   4     x   9
_____  _____  _____  _____  _____
```

A

2	3	5	4	1
x 8	x 2	x 9	x 7	x 6

B

6	10	7	8	5
x 1	x 5	x 4	x 9	x 3

C

10	4	6	2	9
x 1	x 7	x 5	x 9	x 3

D

1	7	3	8	2
x 4	x 6	x 8	x 5	x 2

E

3	6	8	10	4
x 5	x 6	x 8	x 9	x 4

A
```
      7          5          3          4          6
 X    3     X    2     X    5     X    4     X    2
```

B
```
      8          1          9          2         10
 X    7     X    9     X    6     X    1     X    8
```

C
```
      7          3          5          1          9
 X    4     X    6     X    7     X    1     X    9
```

D
```
      6          4          8         10          2
 X    2     X    3     X    8     X    2     X    5
```

E
```
      9          7          3          5         10
 X    8     X    5     X    9     X    7     X   10
```

A

9	4	7	5	1
X 1	X 9	X 3	X 6	X 5

B

10	2	6	3	8
X 8	X 3	X 7	X 4	X 2

C

6	4	8	3	7
X 6	X 8	X 3	X 2	X 5

D

1	9	5	10	2
X 4	X 7	X 6	X 1	X 2

E

6	5	9	1	8
X 8	X 6	X 5	X 9	X 1

A
```
      6           8           1          10           4
  x   2       x   1       x   1       x   1       x   1
  _____   _____   _____   _____   _____
```

B
```
      3           7          11           2          10
  x   2       x   1       x   1       x   2       x   2
  _____   _____   _____   _____   _____
```

C
```
      4           5          10           3          12
  x   2       x   1       x   2       x   1       x   2
  _____   _____   _____   _____   _____
```

D
```
     12           1          11           5           7
  x   1       x   2       x   2       x   2       x   2
  _____   _____   _____   _____   _____
```

E
```
      6           9           8           2           9
  x   1       x   2       x   2       x   1       x   1
  _____   _____   _____   _____   _____
```

A

$$\begin{array}{r} 2 \\ \times\ 4 \\ \hline \end{array}\qquad \begin{array}{r} 9 \\ \times\ 3 \\ \hline \end{array}\qquad \begin{array}{r} 12 \\ \times\ 3 \\ \hline \end{array}\qquad \begin{array}{r} 4 \\ \times\ 3 \\ \hline \end{array}\qquad \begin{array}{r} 11 \\ \times\ 3 \\ \hline \end{array}$$

B

$$\begin{array}{r} 10 \\ \times\ 4 \\ \hline \end{array}\qquad \begin{array}{r} 8 \\ \times\ 3 \\ \hline \end{array}\qquad \begin{array}{r} 5 \\ \times\ 3 \\ \hline \end{array}\qquad \begin{array}{r} 6 \\ \times\ 4 \\ \hline \end{array}\qquad \begin{array}{r} 6 \\ \times\ 4 \\ \hline \end{array}$$

C

$$\begin{array}{r} 12 \\ \times\ 4 \\ \hline \end{array}\qquad \begin{array}{r} 3 \\ \times\ 3 \\ \hline \end{array}\qquad \begin{array}{r} 11 \\ \times\ 4 \\ \hline \end{array}\qquad \begin{array}{r} 7 \\ \times\ 3 \\ \hline \end{array}\qquad \begin{array}{r} 3 \\ \times\ 4 \\ \hline \end{array}$$

D

$$\begin{array}{r} 2 \\ \times\ 3 \\ \hline \end{array}\qquad \begin{array}{r} 7 \\ \times\ 4 \\ \hline \end{array}\qquad \begin{array}{r} 8 \\ \times\ 4 \\ \hline \end{array}\qquad \begin{array}{r} 4 \\ \times\ 4 \\ \hline \end{array}\qquad \begin{array}{r} 5 \\ \times\ 4 \\ \hline \end{array}$$

E

$$\begin{array}{r} 10 \\ \times\ 3 \\ \hline \end{array}\qquad \begin{array}{r} 9 \\ \times\ 4 \\ \hline \end{array}\qquad \begin{array}{r} 1 \\ \times\ 4 \\ \hline \end{array}\qquad \begin{array}{r} 1 \\ \times\ 3 \\ \hline \end{array}\qquad \begin{array}{r} 6 \\ \times\ 3 \\ \hline \end{array}$$

A

$$\begin{array}{r} 6 \\ \times\ 6 \\ \hline \end{array}$$
$$\begin{array}{r} 7 \\ \times\ 5 \\ \hline \end{array}$$
$$\begin{array}{r} 8 \\ \times\ 5 \\ \hline \end{array}$$
$$\begin{array}{r} 6 \\ \times\ 5 \\ \hline \end{array}$$
$$\begin{array}{r} 4 \\ \times\ 5 \\ \hline \end{array}$$

B

$$\begin{array}{r} 3 \\ \times\ 6 \\ \hline \end{array}$$
$$\begin{array}{r} 10 \\ \times\ 5 \\ \hline \end{array}$$
$$\begin{array}{r} 2 \\ \times\ 5 \\ \hline \end{array}$$
$$\begin{array}{r} 7 \\ \times\ 6 \\ \hline \end{array}$$
$$\begin{array}{r} 10 \\ \times\ 6 \\ \hline \end{array}$$

C

$$\begin{array}{r} 4 \\ \times\ 6 \\ \hline \end{array}$$
$$\begin{array}{r} 3 \\ \times\ 5 \\ \hline \end{array}$$
$$\begin{array}{r} 8 \\ \times\ 6 \\ \hline \end{array}$$
$$\begin{array}{r} 12 \\ \times\ 5 \\ \hline \end{array}$$
$$\begin{array}{r} 2 \\ \times\ 6 \\ \hline \end{array}$$

D

$$\begin{array}{r} 1 \\ \times\ 5 \\ \hline \end{array}$$
$$\begin{array}{r} 5 \\ \times\ 6 \\ \hline \end{array}$$
$$\begin{array}{r} 12 \\ \times\ 6 \\ \hline \end{array}$$
$$\begin{array}{r} 11 \\ \times\ 6 \\ \hline \end{array}$$
$$\begin{array}{r} 9 \\ \times\ 6 \\ \hline \end{array}$$

E

$$\begin{array}{r} 11 \\ \times\ 5 \\ \hline \end{array}$$
$$\begin{array}{r} 1 \\ \times\ 6 \\ \hline \end{array}$$
$$\begin{array}{r} 10 \\ \times\ 6 \\ \hline \end{array}$$
$$\begin{array}{r} 5 \\ \times\ 5 \\ \hline \end{array}$$
$$\begin{array}{r} 9 \\ \times\ 5 \\ \hline \end{array}$$

A
$$\begin{array}{r} 7 \\ \times\ 8 \\ \hline \end{array}$$
$$\begin{array}{r} 7 \\ \times\ 7 \\ \hline \end{array}$$
$$\begin{array}{r} 1 \\ \times\ 7 \\ \hline \end{array}$$
$$\begin{array}{r} 6 \\ \times\ 7 \\ \hline \end{array}$$
$$\begin{array}{r} 11 \\ \times\ 7 \\ \hline \end{array}$$

B
$$\begin{array}{r} 6 \\ \times\ 8 \\ \hline \end{array}$$
$$\begin{array}{r} 5 \\ \times\ 7 \\ \hline \end{array}$$
$$\begin{array}{r} 2 \\ \times\ 7 \\ \hline \end{array}$$
$$\begin{array}{r} 3 \\ \times\ 8 \\ \hline \end{array}$$
$$\begin{array}{r} 7 \\ \times\ 8 \\ \hline \end{array}$$

C
$$\begin{array}{r} 2 \\ \times\ 8 \\ \hline \end{array}$$
$$\begin{array}{r} 9 \\ \times\ 7 \\ \hline \end{array}$$
$$\begin{array}{r} 12 \\ \times\ 8 \\ \hline \end{array}$$
$$\begin{array}{r} 12 \\ \times\ 7 \\ \hline \end{array}$$
$$\begin{array}{r} 1 \\ \times\ 8 \\ \hline \end{array}$$

D
$$\begin{array}{r} 8 \\ \times\ 7 \\ \hline \end{array}$$
$$\begin{array}{r} 4 \\ \times\ 8 \\ \hline \end{array}$$
$$\begin{array}{r} 9 \\ \times\ 8 \\ \hline \end{array}$$
$$\begin{array}{r} 5 \\ \times\ 8 \\ \hline \end{array}$$
$$\begin{array}{r} 8 \\ \times\ 8 \\ \hline \end{array}$$

E
$$\begin{array}{r} 4 \\ \times\ 7 \\ \hline \end{array}$$
$$\begin{array}{r} 10 \\ \times\ 8 \\ \hline \end{array}$$
$$\begin{array}{r} 11 \\ \times\ 8 \\ \hline \end{array}$$
$$\begin{array}{r} 3 \\ \times\ 7 \\ \hline \end{array}$$
$$\begin{array}{r} 10 \\ \times\ 7 \\ \hline \end{array}$$

A
$$\begin{array}{r} 5 \\ \times\ 10 \\ \hline \end{array}$$
$$\begin{array}{r} 10 \\ \times\ 9 \\ \hline \end{array}$$
$$\begin{array}{r} 9 \\ \times\ 9 \\ \hline \end{array}$$
$$\begin{array}{r} 11 \\ \times\ 9 \\ \hline \end{array}$$
$$\begin{array}{r} 6 \\ \times\ 9 \\ \hline \end{array}$$

B
$$\begin{array}{r} 3 \\ \times\ 10 \\ \hline \end{array}$$
$$\begin{array}{r} 2 \\ \times\ 9 \\ \hline \end{array}$$
$$\begin{array}{r} 3 \\ \times\ 9 \\ \hline \end{array}$$
$$\begin{array}{r} 6 \\ \times\ 10 \\ \hline \end{array}$$
$$\begin{array}{r} 10 \\ \times\ 10 \\ \hline \end{array}$$

C
$$\begin{array}{r} 10 \\ \times\ 10 \\ \hline \end{array}$$
$$\begin{array}{r} 12 \\ \times\ 9 \\ \hline \end{array}$$
$$\begin{array}{r} 7 \\ \times\ 10 \\ \hline \end{array}$$
$$\begin{array}{r} 4 \\ \times\ 9 \\ \hline \end{array}$$
$$\begin{array}{r} 12 \\ \times\ 10 \\ \hline \end{array}$$

D
$$\begin{array}{r} 7 \\ \times\ 9 \\ \hline \end{array}$$
$$\begin{array}{r} 1 \\ \times\ 10 \\ \hline \end{array}$$
$$\begin{array}{r} 8 \\ \times\ 12 \\ \hline \end{array}$$
$$\begin{array}{r} 11 \\ \times\ 11 \\ \hline \end{array}$$
$$\begin{array}{r} 9 \\ \times\ 10 \\ \hline \end{array}$$

E
$$\begin{array}{r} 1 \\ \times\ 9 \\ \hline \end{array}$$
$$\begin{array}{r} 2 \\ \times\ 10 \\ \hline \end{array}$$
$$\begin{array}{r} 4 \\ \times\ 10 \\ \hline \end{array}$$
$$\begin{array}{r} 8 \\ \times\ 9 \\ \hline \end{array}$$
$$\begin{array}{r} 5 \\ \times\ 9 \\ \hline \end{array}$$

A
```
   12        11        11        11        11
X   7     X   7     X   5     X   2     X   6
```

B
```
   12        11        11        12        12
X  10     X   4     X  11     X   2     X  10
```

C
```
   12        11        12        11        12
X  11     X   1     X   1     X   3     X   5
```

D
```
   11        12        12        12        12
X   9     X   9     X   8     X   3     X   4
```

E
```
   12        12        12        11        11
X  11     X   6     X   8     X  10     X   8
```

A
```
     8           7           4           3          11
 X   8       X   9       X   3       X   3       X   1
```

B
```
     9           5           6           1          12
 X   6       X   6       X   5       X   4       X   2
```

C
```
     2          10           2          10           7
 X   7       X  10       X   9       X  10       X   7
```

D
```
    11           9           4          10          12
 X   4       X   1       X   6       X   2       X   3
```

E
```
     6           1           8           5          10
 X   8       X   5       X   4       X   4       X   3
```

Day: 43

Name:

Date:

Time: :

Score: /25

Rating: ☆☆☆☆☆☆

A
$$10 \times 1$$
$$4 \times 3$$
$$3 \times 5$$
$$2 \times 4$$
$$9 \times 5$$

B
$$7 \times 6$$
$$1 \times 2$$
$$6 \times 8$$
$$11 \times 4$$
$$5 \times 1$$

C
$$8 \times 9$$
$$4 \times 7$$
$$8 \times 2$$
$$6 \times 7$$
$$10 \times 3$$

D
$$3 \times 8$$
$$9 \times 6$$
$$12 \times 6$$
$$1 \times 9$$
$$11 \times 5$$

E
$$2 \times 1$$
$$7 \times 3$$
$$3 \times 4$$
$$10 \times 8$$
$$9 \times 5$$

A
$$
\begin{array}{r} 2 \\ \times\ 6 \\ \hline \end{array}
\qquad
\begin{array}{r} 5 \\ \times\ 9 \\ \hline \end{array}
\qquad
\begin{array}{r} 4 \\ \times\ 1 \\ \hline \end{array}
\qquad
\begin{array}{r} 7 \\ \times\ 3 \\ \hline \end{array}
\qquad
\begin{array}{r} 6 \\ \times\ 5 \\ \hline \end{array}
$$

B
$$
\begin{array}{r} 11 \\ \times\ 3 \\ \hline \end{array}
\qquad
\begin{array}{r} 9 \\ \times\ 4 \\ \hline \end{array}
\qquad
\begin{array}{r} 1 \\ \times\ 8 \\ \hline \end{array}
\qquad
\begin{array}{r} 10 \\ \times\ 8 \\ \hline \end{array}
\qquad
\begin{array}{r} 3 \\ \times\ 2 \\ \hline \end{array}
$$

C
$$
\begin{array}{r} 12 \\ \times\ 7 \\ \hline \end{array}
\qquad
\begin{array}{r} 11 \\ \times\ 5 \\ \hline \end{array}
\qquad
\begin{array}{r} 10 \\ \times\ 1 \\ \hline \end{array}
\qquad
\begin{array}{r} 12 \\ \times\ 4 \\ \hline \end{array}
\qquad
\begin{array}{r} 3 \\ \times\ 1 \\ \hline \end{array}
$$

D
$$
\begin{array}{r} 1 \\ \times\ 8 \\ \hline \end{array}
\qquad
\begin{array}{r} 9 \\ \times\ 7 \\ \hline \end{array}
\qquad
\begin{array}{r} 8 \\ \times\ 2 \\ \hline \end{array}
\qquad
\begin{array}{r} 7 \\ \times\ 9 \\ \hline \end{array}
\qquad
\begin{array}{r} 6 \\ \times\ 3 \\ \hline \end{array}
$$

E
$$
\begin{array}{r} 4 \\ \times\ 6 \\ \hline \end{array}
\qquad
\begin{array}{r} 11 \\ \times\ 10 \\ \hline \end{array}
\qquad
\begin{array}{r} 12 \\ \times\ 4 \\ \hline \end{array}
\qquad
\begin{array}{r} 7 \\ \times\ 5 \\ \hline \end{array}
\qquad
\begin{array}{r} 9 \\ \times\ 5 \\ \hline \end{array}
$$

A

```
    5          10          7          2          3
x   1      x    2      x   4      x   8      x   5
_____    _____    _____    _____    _____
```

B

```
    9           8          4          1          6
x   3      x    5      x   1      x   6      x   5
_____    _____    _____    _____    _____
```

C

```
   12          11          2          4          7
x   7      x    9      x   9      x   6      x   3
_____    _____    _____    _____    _____
```

D

```
    9          12         11          8          1
x   5      x    8      x   6      x   4      x   7
_____    _____    _____    _____    _____
```

E

```
    3          10         11          1          2
x   2      x    3      x   5      x   1      x   9
_____    _____    _____    _____    _____
```

A

```
    3          4          7          1         10
x   4      x   7      x   2      x   1      x   9
```

B

```
   12         10          8          5         11
x  11      x   3      x   6      x   5      x   9
```

C

```
   10          2         10          5         12
x   6      x   8      x   1      x   2      x   5
```

D

```
    4         11          1          8          7
x   7      x   2      x   8      x   7      x   9
```

E

```
    9          3         12          5          1
x   9      x   6      x   4      x   3      x   3
```

A
$$\begin{array}{r} 11 \\ \times\ \ 2 \\ \hline \end{array}$$
$$\begin{array}{r} 4 \\ \times\ \ 1 \\ \hline \end{array}$$
$$\begin{array}{r} 12 \\ \times\ \ 7 \\ \hline \end{array}$$
$$\begin{array}{r} 5 \\ \times\ \ 4 \\ \hline \end{array}$$
$$\begin{array}{r} 3 \\ \times\ \ 5 \\ \hline \end{array}$$

B
$$\begin{array}{r} 1 \\ \times\ \ 3 \\ \hline \end{array}$$
$$\begin{array}{r} 7 \\ \times\ \ 6 \\ \hline \end{array}$$
$$\begin{array}{r} 6 \\ \times\ \ 8 \\ \hline \end{array}$$
$$\begin{array}{r} 8 \\ \times\ \ 8 \\ \hline \end{array}$$
$$\begin{array}{r} 10 \\ \times\ \ 5 \\ \hline \end{array}$$

C
$$\begin{array}{r} 9 \\ \times\ \ 5 \\ \hline \end{array}$$
$$\begin{array}{r} 10 \\ \times\ \ 5 \\ \hline \end{array}$$
$$\begin{array}{r} 7 \\ \times\ \ 3 \\ \hline \end{array}$$
$$\begin{array}{r} 2 \\ \times\ \ 2 \\ \hline \end{array}$$
$$\begin{array}{r} 11 \\ \times\ \ 3 \\ \hline \end{array}$$

D
$$\begin{array}{r} 12 \\ \times\ \ 1 \\ \hline \end{array}$$
$$\begin{array}{r} 5 \\ \times\ \ 7 \\ \hline \end{array}$$
$$\begin{array}{r} 4 \\ \times\ \ 4 \\ \hline \end{array}$$
$$\begin{array}{r} 3 \\ \times\ \ 2 \\ \hline \end{array}$$
$$\begin{array}{r} 1 \\ \times\ \ 9 \\ \hline \end{array}$$

E
$$\begin{array}{r} 8 \\ \times\ \ 9 \\ \hline \end{array}$$
$$\begin{array}{r} 10 \\ \times\ \ 8 \\ \hline \end{array}$$
$$\begin{array}{r} 9 \\ \times\ \ 6 \\ \hline \end{array}$$
$$\begin{array}{r} 2 \\ \times\ \ 3 \\ \hline \end{array}$$
$$\begin{array}{r} 8 \\ \times\ \ 4 \\ \hline \end{array}$$

A

$$\begin{array}{r} 12 \\ \times\ 3 \\ \hline \end{array}\qquad \begin{array}{r} 11 \\ \times\ 4 \\ \hline \end{array}\qquad \begin{array}{r} 8 \\ \times\ 3 \\ \hline \end{array}\qquad \begin{array}{r} 5 \\ \times\ 9 \\ \hline \end{array}\qquad \begin{array}{r} 9 \\ \times\ 5 \\ \hline \end{array}$$

B

$$\begin{array}{r} 6 \\ \times\ 7 \\ \hline \end{array}\qquad \begin{array}{r} 1 \\ \times\ 7 \\ \hline \end{array}\qquad \begin{array}{r} 7 \\ \times\ 8 \\ \hline \end{array}\qquad \begin{array}{r} 10 \\ \times\ 6 \\ \hline \end{array}\qquad \begin{array}{r} 3 \\ \times\ 1 \\ \hline \end{array}$$

C

$$\begin{array}{r} 2 \\ \times\ 2 \\ \hline \end{array}\qquad \begin{array}{r} 4 \\ \times\ 5 \\ \hline \end{array}\qquad \begin{array}{r} 8 \\ \times\ 4 \\ \hline \end{array}\qquad \begin{array}{r} 3 \\ \times\ 9 \\ \hline \end{array}\qquad \begin{array}{r} 12 \\ \times\ 1 \\ \hline \end{array}$$

D

$$\begin{array}{r} 1 \\ \times\ 4 \\ \hline \end{array}\qquad \begin{array}{r} 2 \\ \times\ 5 \\ \hline \end{array}\qquad \begin{array}{r} 6 \\ \times\ 2 \\ \hline \end{array}\qquad \begin{array}{r} 10 \\ \times\ 7 \\ \hline \end{array}\qquad \begin{array}{r} 11 \\ \times\ 7 \\ \hline \end{array}$$

E

$$\begin{array}{r} 4 \\ \times\ 6 \\ \hline \end{array}\qquad \begin{array}{r} 7 \\ \times\ 3 \\ \hline \end{array}\qquad \begin{array}{r} 5 \\ \times\ 8 \\ \hline \end{array}\qquad \begin{array}{r} 12 \\ \times\ 2 \\ \hline \end{array}\qquad \begin{array}{r} 10 \\ \times\ 10 \\ \hline \end{array}$$

A

$$\begin{array}{r} 8 \\ \times\ 2 \\ \hline \end{array}\qquad \begin{array}{r} 7 \\ \times\ 4 \\ \hline \end{array}\qquad \begin{array}{r} 6 \\ \times\ 3 \\ \hline \end{array}\qquad \begin{array}{r} 5 \\ \times\ 3 \\ \hline \end{array}\qquad \begin{array}{r} 1 \\ \times\ 8 \\ \hline \end{array}$$

B

$$\begin{array}{r} 4 \\ \times\ 5 \\ \hline \end{array}\qquad \begin{array}{r} 10 \\ \times\ 2 \\ \hline \end{array}\qquad \begin{array}{r} 3 \\ \times\ 9 \\ \hline \end{array}\qquad \begin{array}{r} 12 \\ \times\ 9 \\ \hline \end{array}\qquad \begin{array}{r} 9 \\ \times\ 6 \\ \hline \end{array}$$

C

$$\begin{array}{r} 10 \\ \times\ 7 \\ \hline \end{array}\qquad \begin{array}{r} 1 \\ \times\ 1 \\ \hline \end{array}\qquad \begin{array}{r} 9 \\ \times\ 9 \\ \hline \end{array}\qquad \begin{array}{r} 4 \\ \times\ 6 \\ \hline \end{array}\qquad \begin{array}{r} 12 \\ \times\ 4 \\ \hline \end{array}$$

D

$$\begin{array}{r} 11 \\ \times\ 7 \\ \hline \end{array}\qquad \begin{array}{r} 5 \\ \times\ 7 \\ \hline \end{array}\qquad \begin{array}{r} 8 \\ \times\ 2 \\ \hline \end{array}\qquad \begin{array}{r} 6 \\ \times\ 7 \\ \hline \end{array}\qquad \begin{array}{r} 3 \\ \times\ 3 \\ \hline \end{array}$$

E

$$\begin{array}{r} 10 \\ \times\ 8 \\ \hline \end{array}\qquad \begin{array}{r} 11 \\ \times\ 10 \\ \hline \end{array}\qquad \begin{array}{r} 11 \\ \times\ 1 \\ \hline \end{array}\qquad \begin{array}{r} 7 \\ \times\ 5 \\ \hline \end{array}\qquad \begin{array}{r} 8 \\ \times\ 4 \\ \hline \end{array}$$

Day: 50

Date:

Score: /25

Name:

Time: :

Rating: ☆☆☆☆☆☆

A
$$\begin{array}{r} 12 \\ \times\ 10 \\ \hline \end{array}$$
$$\begin{array}{r} 11 \\ \times\ 9 \\ \hline \end{array}$$
$$\begin{array}{r} 12 \\ \times\ 1 \\ \hline \end{array}$$
$$\begin{array}{r} 3 \\ \times\ 4 \\ \hline \end{array}$$
$$\begin{array}{r} 4 \\ \times\ 7 \\ \hline \end{array}$$

B
$$\begin{array}{r} 2 \\ \times\ 5 \\ \hline \end{array}$$
$$\begin{array}{r} 5 \\ \times\ 7 \\ \hline \end{array}$$
$$\begin{array}{r} 8 \\ \times\ 2 \\ \hline \end{array}$$
$$\begin{array}{r} 6 \\ \times\ 6 \\ \hline \end{array}$$
$$\begin{array}{r} 9 \\ \times\ 3 \\ \hline \end{array}$$

C
$$\begin{array}{r} 7 \\ \times\ 8 \\ \hline \end{array}$$
$$\begin{array}{r} 11 \\ \times\ 4 \\ \hline \end{array}$$
$$\begin{array}{r} 5 \\ \times\ 2 \\ \hline \end{array}$$
$$\begin{array}{r} 1 \\ \times\ 2 \\ \hline \end{array}$$
$$\begin{array}{r} 7 \\ \times\ 9 \\ \hline \end{array}$$

D
$$\begin{array}{r} 12 \\ \times\ 4 \\ \hline \end{array}$$
$$\begin{array}{r} 6 \\ \times\ 1 \\ \hline \end{array}$$
$$\begin{array}{r} 2 \\ \times\ 8 \\ \hline \end{array}$$
$$\begin{array}{r} 10 \\ \times\ 7 \\ \hline \end{array}$$
$$\begin{array}{r} 8 \\ \times\ 4 \\ \hline \end{array}$$

E
$$\begin{array}{r} 11 \\ \times\ 5 \\ \hline \end{array}$$
$$\begin{array}{r} 12 \\ \times\ 6 \\ \hline \end{array}$$
$$\begin{array}{r} 9 \\ \times\ 3 \\ \hline \end{array}$$
$$\begin{array}{r} 11 \\ \times\ 10 \\ \hline \end{array}$$
$$\begin{array}{r} 10 \\ \times\ 2 \\ \hline \end{array}$$

A

$$\begin{array}{r} 4 \\ \times\ 1 \\ \hline \end{array}\qquad \begin{array}{r} 10 \\ \times\ 3 \\ \hline \end{array}\qquad \begin{array}{r} 6 \\ \times\ 7 \\ \hline \end{array}\qquad \begin{array}{r} 9 \\ \times\ 1 \\ \hline \end{array}\qquad \begin{array}{r} 7 \\ \times\ 5 \\ \hline \end{array}$$

B

$$\begin{array}{r} 5 \\ \times\ 6 \\ \hline \end{array}\qquad \begin{array}{r} 10 \\ \times\ 5 \\ \hline \end{array}\qquad \begin{array}{r} 11 \\ \times\ 3 \\ \hline \end{array}\qquad \begin{array}{r} 2 \\ \times\ 2 \\ \hline \end{array}\qquad \begin{array}{r} 1 \\ \times\ 4 \\ \hline \end{array}$$

C

$$\begin{array}{r} 12 \\ \times\ 10 \\ \hline \end{array}\qquad \begin{array}{r} 8 \\ \times\ 9 \\ \hline \end{array}\qquad \begin{array}{r} 8 \\ \times\ 3 \\ \hline \end{array}\qquad \begin{array}{r} 6 \\ \times\ 2 \\ \hline \end{array}\qquad \begin{array}{r} 2 \\ \times\ 3 \\ \hline \end{array}$$

D

$$\begin{array}{r} 11 \\ \times\ 6 \\ \hline \end{array}\qquad \begin{array}{r} 4 \\ \times\ 1 \\ \hline \end{array}\qquad \begin{array}{r} 3 \\ \times\ 8 \\ \hline \end{array}\qquad \begin{array}{r} 10 \\ \times\ 5 \\ \hline \end{array}\qquad \begin{array}{r} 7 \\ \times\ 7 \\ \hline \end{array}$$

E

$$\begin{array}{r} 12 \\ \times\ 9 \\ \hline \end{array}\qquad \begin{array}{r} 9 \\ \times\ 4 \\ \hline \end{array}\qquad \begin{array}{r} 5 \\ \times\ 2 \\ \hline \end{array}\qquad \begin{array}{r} 11 \\ \times\ 1 \\ \hline \end{array}\qquad \begin{array}{r} 12 \\ \times\ 10 \\ \hline \end{array}$$

A

11	20	19	16	10
X 6	X 9	X 1	X 2	X 4

B

13	12	18	15	14
X 7	X 4	X 5	X 2	X 1

C

17	15	17	18	16
X 9	X 2	X 9	X 8	X 0

D

19	13	20	14	11
X 5	X 4	X 1	X 0	X 2

E

10	12	10	15	20
X 4	X 8	X 1	X 3	X 7

A

$$\begin{array}{r} 15 \\ \times\ \ 0 \\ \hline \end{array}$$
$$\begin{array}{r} 20 \\ \times\ \ 6 \\ \hline \end{array}$$
$$\begin{array}{r} 16 \\ \times\ \ 7 \\ \hline \end{array}$$
$$\begin{array}{r} 13 \\ \times\ \ 2 \\ \hline \end{array}$$
$$\begin{array}{r} 12 \\ \times\ \ 1 \\ \hline \end{array}$$

B

$$\begin{array}{r} 17 \\ \times\ \ 9 \\ \hline \end{array}$$
$$\begin{array}{r} 18 \\ \times\ \ 4 \\ \hline \end{array}$$
$$\begin{array}{r} 10 \\ \times\ \ 5 \\ \hline \end{array}$$
$$\begin{array}{r} 14 \\ \times\ \ 3 \\ \hline \end{array}$$
$$\begin{array}{r} 19 \\ \times\ \ 8 \\ \hline \end{array}$$

C

$$\begin{array}{r} 11 \\ \times\ \ 3 \\ \hline \end{array}$$
$$\begin{array}{r} 10 \\ \times\ \ 0 \\ \hline \end{array}$$
$$\begin{array}{r} 16 \\ \times\ \ 1 \\ \hline \end{array}$$
$$\begin{array}{r} 15 \\ \times\ \ 8 \\ \hline \end{array}$$
$$\begin{array}{r} 17 \\ \times\ \ 2 \\ \hline \end{array}$$

D

$$\begin{array}{r} 18 \\ \times\ \ 6 \\ \hline \end{array}$$
$$\begin{array}{r} 13 \\ \times\ \ 5 \\ \hline \end{array}$$
$$\begin{array}{r} 14 \\ \times\ \ 4 \\ \hline \end{array}$$
$$\begin{array}{r} 20 \\ \times\ \ 7 \\ \hline \end{array}$$
$$\begin{array}{r} 11 \\ \times\ \ 9 \\ \hline \end{array}$$

E

$$\begin{array}{r} 19 \\ \times\ \ 8 \\ \hline \end{array}$$
$$\begin{array}{r} 18 \\ \times\ \ 7 \\ \hline \end{array}$$
$$\begin{array}{r} 16 \\ \times\ \ 4 \\ \hline \end{array}$$
$$\begin{array}{r} 13 \\ \times\ \ 0 \\ \hline \end{array}$$
$$\begin{array}{r} 11 \\ \times\ \ 5 \\ \hline \end{array}$$

A

13	15	11	19	12
X 0	X 7	X 9	X 1	X 3

B

18	16	20	14	17
X 5	X 2	X 8	X 4	X 6

C

10	15	14	17	10
X 8	X 9	X 1	X 0	X 3

D

18	20	13	16	19
X 2	X 6	X 4	X 7	X 5

E

11	14	19	10	20
X 9	X 4	X 6	X 2	X 7

A

10	20	16	14	15
X 9	X 0	X 7	X 4	X 1

B

18	19	12	17	11
X 8	X 6	X 3	X 2	X 5

C

13	18	20	10	19
X 4	X 7	X 1	X 9	X 6

D

13	15	17	11	16
X 4	X 0	X 3	X 5	X 8

E

12	18	16	10	17
X 2	X 0	X 6	X 3	X 7

A
| 16 | 20 | 13 | 19 | 12 |
| X   3 | X   9 | X   4 | X   0 | X   5 |

B
| 17 | 15 | 14 | 18 | 11 |
| X   1 | X   2 | X   8 | X   7 | X   6 |

C
| 10 | 12 | 18 | 20 | 17 |
| X   8 | X   6 | X   0 | X   7 | X   3 |

D
| 10 | 16 | 13 | 15 | 14 |
| X   4 | X   2 | X   9 | X   1 | X   5 |

E
| 11 | 14 | 10 | 19 | 15 |
| X   4 | X   7 | X   4 | X   2 | X   5 |

A

$$\begin{array}{r} 14 \\ \times\ 8 \\ \hline \end{array}\qquad \begin{array}{r} 15 \\ \times\ 2 \\ \hline \end{array}\qquad \begin{array}{r} 13 \\ \times\ 4 \\ \hline \end{array}\qquad \begin{array}{r} 17 \\ \times\ 3 \\ \hline \end{array}\qquad \begin{array}{r} 16 \\ \times\ 9 \\ \hline \end{array}$$

B

$$\begin{array}{r} 12 \\ \times\ 0 \\ \hline \end{array}\qquad \begin{array}{r} 18 \\ \times\ 7 \\ \hline \end{array}\qquad \begin{array}{r} 20 \\ \times\ 6 \\ \hline \end{array}\qquad \begin{array}{r} 11 \\ \times\ 1 \\ \hline \end{array}\qquad \begin{array}{r} 19 \\ \times\ 5 \\ \hline \end{array}$$

C

$$\begin{array}{r} 10 \\ \times\ 3 \\ \hline \end{array}\qquad \begin{array}{r} 20 \\ \times\ 2 \\ \hline \end{array}\qquad \begin{array}{r} 12 \\ \times\ 5 \\ \hline \end{array}\qquad \begin{array}{r} 19 \\ \times\ 6 \\ \hline \end{array}\qquad \begin{array}{r} 18 \\ \times\ 4 \\ \hline \end{array}$$

D

$$\begin{array}{r} 15 \\ \times\ 1 \\ \hline \end{array}\qquad \begin{array}{r} 16 \\ \times\ 8 \\ \hline \end{array}\qquad \begin{array}{r} 10 \\ \times\ 7 \\ \hline \end{array}\qquad \begin{array}{r} 14 \\ \times\ 0 \\ \hline \end{array}\qquad \begin{array}{r} 17 \\ \times\ 9 \\ \hline \end{array}$$

E

$$\begin{array}{r} 13 \\ \times\ 1 \\ \hline \end{array}\qquad \begin{array}{r} 18 \\ \times\ 6 \\ \hline \end{array}\qquad \begin{array}{r} 12 \\ \times\ 7 \\ \hline \end{array}\qquad \begin{array}{r} 14 \\ \times\ 9 \\ \hline \end{array}\qquad \begin{array}{r} 16 \\ \times\ 3 \\ \hline \end{array}$$

A
```
     20          18          13          19          14
x     5       x     8     x     3     x     7     x     6
```

B
```
     17          15          10          11          16
x     2       x     4     x     1     x     0     x     9
```

C
```
     12          17          16          18          12
x     0       x     7     x     4     x     5     x     9
```

D
```
     11          19          13          15          10
x     3       x     8     x     2     x     1     x     6
```

E
```
     20          17          15          18          12
x     4       x     0     x     7     x     8     x     9
```

A

| 20
× 7 | 15
× 6 | 17
× 9 | 18
× 8 | 16
× 1 |

B

| 11
× 5 | 14
× 0 | 12
× 2 | 10
× 4 | 13
× 3 |

C

| 19
× 0 | 12
× 8 | 19
× 5 | 16
× 1 | 11
× 3 |

D

| 18
× 7 | 10
× 2 | 14
× 9 | 13
× 4 | 17
× 6 |

E

| 15
× 5 | 13
× 3 | 15
× 6 | 11
× 9 | 18
× 2 |

A
$$\begin{array}{r} 16 \\ \times\ 4 \\ \hline \end{array}$$
$$\begin{array}{r} 15 \\ \times\ 3 \\ \hline \end{array}$$
$$\begin{array}{r} 20 \\ \times\ 5 \\ \hline \end{array}$$
$$\begin{array}{r} 19 \\ \times\ 6 \\ \hline \end{array}$$
$$\begin{array}{r} 10 \\ \times\ 9 \\ \hline \end{array}$$

B
$$\begin{array}{r} 11 \\ \times\ 8 \\ \hline \end{array}$$
$$\begin{array}{r} 13 \\ \times\ 7 \\ \hline \end{array}$$
$$\begin{array}{r} 18 \\ \times\ 2 \\ \hline \end{array}$$
$$\begin{array}{r} 12 \\ \times\ 0 \\ \hline \end{array}$$
$$\begin{array}{r} 14 \\ \times\ 1 \\ \hline \end{array}$$

C
$$\begin{array}{r} 17 \\ \times\ 5 \\ \hline \end{array}$$
$$\begin{array}{r} 15 \\ \times\ 9 \\ \hline \end{array}$$
$$\begin{array}{r} 11 \\ \times\ 3 \\ \hline \end{array}$$
$$\begin{array}{r} 18 \\ \times\ 0 \\ \hline \end{array}$$
$$\begin{array}{r} 20 \\ \times\ 1 \\ \hline \end{array}$$

D
$$\begin{array}{r} 12 \\ \times\ 4 \\ \hline \end{array}$$
$$\begin{array}{r} 14 \\ \times\ 7 \\ \hline \end{array}$$
$$\begin{array}{r} 17 \\ \times\ 6 \\ \hline \end{array}$$
$$\begin{array}{r} 19 \\ \times\ 2 \\ \hline \end{array}$$
$$\begin{array}{r} 10 \\ \times\ 8 \\ \hline \end{array}$$

E
$$\begin{array}{r} 16 \\ \times\ 3 \\ \hline \end{array}$$
$$\begin{array}{r} 14 \\ \times\ 5 \\ \hline \end{array}$$
$$\begin{array}{r} 17 \\ \times\ 1 \\ \hline \end{array}$$
$$\begin{array}{r} 10 \\ \times\ 6 \\ \hline \end{array}$$
$$\begin{array}{r} 11 \\ \times\ 7 \\ \hline \end{array}$$

A

$$
\begin{array}{r} 18 \\ \times\ 8 \\ \hline \end{array}
\qquad
\begin{array}{r} 19 \\ \times\ 5 \\ \hline \end{array}
\qquad
\begin{array}{r} 13 \\ \times\ 9 \\ \hline \end{array}
\qquad
\begin{array}{r} 10 \\ \times\ 0 \\ \hline \end{array}
\qquad
\begin{array}{r} 15 \\ \times\ 4 \\ \hline \end{array}
$$

B

$$
\begin{array}{r} 14 \\ \times\ 6 \\ \hline \end{array}
\qquad
\begin{array}{r} 11 \\ \times\ 3 \\ \hline \end{array}
\qquad
\begin{array}{r} 12 \\ \times\ 7 \\ \hline \end{array}
\qquad
\begin{array}{r} 17 \\ \times\ 2 \\ \hline \end{array}
\qquad
\begin{array}{r} 20 \\ \times\ 1 \\ \hline \end{array}
$$

C

$$
\begin{array}{r} 16 \\ \times\ 6 \\ \hline \end{array}
\qquad
\begin{array}{r} 19 \\ \times\ 8 \\ \hline \end{array}
\qquad
\begin{array}{r} 17 \\ \times\ 0 \\ \hline \end{array}
\qquad
\begin{array}{r} 13 \\ \times\ 9 \\ \hline \end{array}
\qquad
\begin{array}{r} 14 \\ \times\ 5 \\ \hline \end{array}
$$

D

$$
\begin{array}{r} 15 \\ \times\ 2 \\ \hline \end{array}
\qquad
\begin{array}{r} 16 \\ \times\ 4 \\ \hline \end{array}
\qquad
\begin{array}{r} 20 \\ \times\ 3 \\ \hline \end{array}
\qquad
\begin{array}{r} 11 \\ \times\ 1 \\ \hline \end{array}
\qquad
\begin{array}{r} 10 \\ \times\ 7 \\ \hline \end{array}
$$

E

$$
\begin{array}{r} 12 \\ \times\ 3 \\ \hline \end{array}
\qquad
\begin{array}{r} 10 \\ \times\ 5 \\ \hline \end{array}
\qquad
\begin{array}{r} 18 \\ \times\ 7 \\ \hline \end{array}
\qquad
\begin{array}{r} 17 \\ \times\ 4 \\ \hline \end{array}
\qquad
\begin{array}{r} 16 \\ \times\ 6 \\ \hline \end{array}
$$

A
```
   93        28        82        48        22
X   9     X   5     X   2     X   8     X   6
```

B
```
   74        96        19        60        84
X   7     X   3     X   4     X   1     X   2
```

C
```
   27        68        38        30        63
X   4     X   9     X   7     X   3     X   1
```

D
```
   66        83        33        98        23
X   6     X   8     X   5     X   7     X   3
```

E
```
   43        77        99        12        67
X   2     X   5     X   6     X   1     X   8
```

Day: 63

Name:

Date:

Time: :

Score: /25

Rating: ☆☆☆☆☆☆

A
```
    66
x    7
```
```
    82
x    5
```
```
    86
x    4
```
```
    15
x    6
```
```
    62
x    2
```

B
```
    16
x    3
```
```
    60
x    9
```
```
    51
x    8
```
```
    96
x    1
```
```
    37
x    2
```

C
```
    71
x    5
```
```
    13
x    9
```
```
    64
x    1
```
```
    58
x    4
```
```
    89
x    6
```

D
```
    28
x    7
```
```
    57
x    8
```
```
    38
x    3
```
```
    20
x    1
```
```
    36
x    2
```

E
```
    26
x    8
```
```
    23
x    4
```
```
    42
x    6
```
```
    56
x    7
```
```
    27
x    9
```

A

23	85	84	46	27
X 2	X 5	X 1	X 6	X 3

B

73	60	91	92	83
X 8	X 4	X 9	X 7	X 5

C

67	52	10	99	80
X 3	X 8	X 7	X 2	X 1

D

56	34	59	30	13
X 6	X 4	X 9	X 7	X 9

E

68	45	21	35	51
X 6	X 1	X 5	X 3	X 4

A

28	97	71	26	58
X 9	X 8	X 3	X 7	X 6

B

27	50	43	24	89
X 4	X 1	X 2	X 5	X 8

C

52	19	93	63	18
X 4	X 1	X 6	X 3	X 9

D

59	84	75	46	90
X 5	X 2	X 7	X 8	X 5

E

31	60	12	82	23
X 3	X 1	X 6	X 4	X 9

A

$$\begin{array}{r} 72 \\ \times\ \ 5 \\ \hline \end{array}$$
$$\begin{array}{r} 96 \\ \times\ \ 6 \\ \hline \end{array}$$
$$\begin{array}{r} 73 \\ \times\ \ 9 \\ \hline \end{array}$$
$$\begin{array}{r} 50 \\ \times\ \ 3 \\ \hline \end{array}$$
$$\begin{array}{r} 36 \\ \times\ \ 2 \\ \hline \end{array}$$

B

$$\begin{array}{r} 18 \\ \times\ \ 7 \\ \hline \end{array}$$
$$\begin{array}{r} 81 \\ \times\ \ 4 \\ \hline \end{array}$$
$$\begin{array}{r} 15 \\ \times\ \ 8 \\ \hline \end{array}$$
$$\begin{array}{r} 71 \\ \times\ \ 1 \\ \hline \end{array}$$
$$\begin{array}{r} 54 \\ \times\ \ 7 \\ \hline \end{array}$$

C

$$\begin{array}{r} 46 \\ \times\ \ 3 \\ \hline \end{array}$$
$$\begin{array}{r} 34 \\ \times\ \ 6 \\ \hline \end{array}$$
$$\begin{array}{r} 35 \\ \times\ \ 1 \\ \hline \end{array}$$
$$\begin{array}{r} 10 \\ \times\ \ 8 \\ \hline \end{array}$$
$$\begin{array}{r} 98 \\ \times\ \ 9 \\ \hline \end{array}$$

D

$$\begin{array}{r} 83 \\ \times\ \ 2 \\ \hline \end{array}$$
$$\begin{array}{r} 42 \\ \times\ \ 4 \\ \hline \end{array}$$
$$\begin{array}{r} 84 \\ \times\ \ 5 \\ \hline \end{array}$$
$$\begin{array}{r} 14 \\ \times\ \ 7 \\ \hline \end{array}$$
$$\begin{array}{r} 99 \\ \times\ \ 6 \\ \hline \end{array}$$

E

$$\begin{array}{r} 58 \\ \times\ \ 5 \\ \hline \end{array}$$
$$\begin{array}{r} 45 \\ \times\ \ 2 \\ \hline \end{array}$$
$$\begin{array}{r} 87 \\ \times\ \ 8 \\ \hline \end{array}$$
$$\begin{array}{r} 53 \\ \times\ \ 1 \\ \hline \end{array}$$
$$\begin{array}{r} 78 \\ \times\ \ 4 \\ \hline \end{array}$$

A

82 × 1	35 × 5	52 × 6	15 × 2	64 × 3

B

31 × 7	85 × 9	59 × 4	60 × 8	37 × 6

C

63 × 4	45 × 9	87 × 8	65 × 5	69 × 2

D

40 × 7	53 × 3	44 × 1	86 × 1	29 × 2

E

83 × 7	34 × 5	57 × 4	70 × 6	43 × 3

A

31	32	90	11	85
× 2	× 4	× 3	× 7	× 9

B

33	56	47	88	94
× 6	× 8	× 5	× 1	× 3

C

98	29	80	72	27
× 1	× 6	× 9	× 8	× 2

D

83	23	61	19	81
× 4	× 7	× 5	× 3	× 1

E

53	68	21	64	59
× 9	× 6	× 2	× 4	× 5

A

65	43	80	60	15
X 5	X 3	X 6	X 2	X 4

B

98	33	69	41	83
X 1	X 8	X 9	X 7	X 3

C

87	90	99	44	35
X 5	X 7	X 2	X 8	X 9

D

86	96	73	40	57
X 4	X 6	X 1	X 7	X 3

E

49	55	24	27	70
X 6	X 5	X 1	X 2	X 4

A

38	84	13	64	97
X 2	X 4	X 5	X 9	X 6

B

47	46	30	28	21
X 7	X 3	X 8	X 5	X 3

C

63	27	77	43	85
X 9	X 7	X 2	X 6	X 4

D

57	74	40	18	83
X 8	X 4	X 3	X 2	X 5

E

56	79	66	16	26
X 9	X 6	X 7	X 8	X 1

A
$$\begin{array}{r}34\\ \times\ 6\\ \hline\end{array}$$
$$\begin{array}{r}47\\ \times\ 2\\ \hline\end{array}$$
$$\begin{array}{r}97\\ \times\ 8\\ \hline\end{array}$$
$$\begin{array}{r}44\\ \times\ 7\\ \hline\end{array}$$
$$\begin{array}{r}37\\ \times\ 5\\ \hline\end{array}$$

B
$$\begin{array}{r}51\\ \times\ 3\\ \hline\end{array}$$
$$\begin{array}{r}79\\ \times\ 9\\ \hline\end{array}$$
$$\begin{array}{r}32\\ \times\ 4\\ \hline\end{array}$$
$$\begin{array}{r}91\\ \times\ 7\\ \hline\end{array}$$
$$\begin{array}{r}62\\ \times\ 4\\ \hline\end{array}$$

C
$$\begin{array}{r}53\\ \times\ 3\\ \hline\end{array}$$
$$\begin{array}{r}90\\ \times\ 2\\ \hline\end{array}$$
$$\begin{array}{r}89\\ \times\ 8\\ \hline\end{array}$$
$$\begin{array}{r}76\\ \times\ 5\\ \hline\end{array}$$
$$\begin{array}{r}86\\ \times\ 9\\ \hline\end{array}$$

D
$$\begin{array}{r}45\\ \times\ 6\\ \hline\end{array}$$
$$\begin{array}{r}36\\ \times\ 2\\ \hline\end{array}$$
$$\begin{array}{r}38\\ \times\ 8\\ \hline\end{array}$$
$$\begin{array}{r}71\\ \times\ 9\\ \hline\end{array}$$
$$\begin{array}{r}43\\ \times\ 7\\ \hline\end{array}$$

E
$$\begin{array}{r}66\\ \times\ 6\\ \hline\end{array}$$
$$\begin{array}{r}99\\ \times\ 3\\ \hline\end{array}$$
$$\begin{array}{r}98\\ \times\ 5\\ \hline\end{array}$$
$$\begin{array}{r}75\\ \times\ 4\\ \hline\end{array}$$
$$\begin{array}{r}12\\ \times\ 5\\ \hline\end{array}$$

A

$$
\begin{array}{r} 96 \\ \times\ 2 \\ \hline \end{array}
\qquad
\begin{array}{r} 34 \\ \times\ 4 \\ \hline \end{array}
\qquad
\begin{array}{r} 65 \\ \times\ 5 \\ \hline \end{array}
\qquad
\begin{array}{r} 95 \\ \times\ 9 \\ \hline \end{array}
\qquad
\begin{array}{r} 89 \\ \times\ 6 \\ \hline \end{array}
$$

B

$$
\begin{array}{r} 40 \\ \times\ 6 \\ \hline \end{array}
\qquad
\begin{array}{r} 37 \\ \times\ 2 \\ \hline \end{array}
\qquad
\begin{array}{r} 77 \\ \times\ 3 \\ \hline \end{array}
\qquad
\begin{array}{r} 68 \\ \times\ 7 \\ \hline \end{array}
\qquad
\begin{array}{r} 55 \\ \times\ 5 \\ \hline \end{array}
$$

C

$$
\begin{array}{r} 73 \\ \times\ 9 \\ \hline \end{array}
\qquad
\begin{array}{r} 36 \\ \times\ 1 \\ \hline \end{array}
\qquad
\begin{array}{r} 33 \\ \times\ 8 \\ \hline \end{array}
\qquad
\begin{array}{r} 50 \\ \times\ 4 \\ \hline \end{array}
\qquad
\begin{array}{r} 61 \\ \times\ 4 \\ \hline \end{array}
$$

D

$$
\begin{array}{r} 47 \\ \times\ 8 \\ \hline \end{array}
\qquad
\begin{array}{r} 98 \\ \times\ 4 \\ \hline \end{array}
\qquad
\begin{array}{r} 70 \\ \times\ 3 \\ \hline \end{array}
\qquad
\begin{array}{r} 91 \\ \times\ 2 \\ \hline \end{array}
\qquad
\begin{array}{r} 12 \\ \times\ 5 \\ \hline \end{array}
$$

E

$$
\begin{array}{r} 32 \\ \times\ 9 \\ \hline \end{array}
\qquad
\begin{array}{r} 87 \\ \times\ 6 \\ \hline \end{array}
\qquad
\begin{array}{r} 48 \\ \times\ 7 \\ \hline \end{array}
\qquad
\begin{array}{r} 24 \\ \times\ 8 \\ \hline \end{array}
\qquad
\begin{array}{r} 22 \\ \times\ 1 \\ \hline \end{array}
$$

A

88	79	34	26	30
x 5	x 3	x 6	x 2	x 4

B

47	95	14	78	68
x 1	x 8	x 9	x 7	x 3

C

20	58	72	93	37
x 5	x 7	x 2	x 8	x 9

D

55	22	27	83	81
x 4	x 6	x 1	x 7	x 3

E

40	15	11	91	97
x 6	x 5	x 1	x 2	x 4

A

20	50	23	79	43
X 2	X 4	X 3	X 7	X 9

B

89	52	57	34	71
X 6	X 8	X 5	X 2	X 7

C

81	18	26	86	82
X 8	X 3	X 5	X 1	X 9

D

84	61	80	21	28
X 6	X 4	X 5	X 3	X 1

E

54	62	58	30	65
X 9	X 6	X 2	X 4	X 5

A
```
    47        27        28        84        19
x    6    x    8    x    6    x    9    x    5
```

B
```
    63        79        11        73        72
x    2    x    4    x    3    x    1    x    7
```

C
```
    88        58        25        46        52
x    1    x    6    x    5    x    5    x    9
```

D
```
    95        78        16        77        41
x    6    x    2    x    8    x    9    x    7
```

E
```
    54        50        10        69        13
x    6    x    3    x    5    x    4    x    5
```

A

87	41	80	37	12
X 1	X 5	X 6	X 2	X 3

B

25	86	64	23	22
X 7	X 9	X 4	X 8	X 6

C

73	99	20	50	91
X 4	X 9	X 8	X 5	X 2

D

70	39	10	47	60
X 7	X 3	X 1	X 1	X 2

E

52	19	69	75	11
X 7	X 5	X 4	X 6	X 3

A

91	38	97	44	21
× 3	× 5	× 4	× 2	× 1

B

15	92	23	31	28
× 6	× 9	× 8	× 7	× 3

C

50	62	32	88	83
× 7	× 4	× 8	× 9	× 2

D

27	35	43	29	34
× 6	× 1	× 5	× 5	× 7

E

98	24	94	58	71
× 1	× 6	× 2	× 3	× 4

Day: 78

Date:

Score: /25

Name:

Time: :

Rating: ☆☆☆☆☆☆

A
$$77 \times 6$$
$$12 \times 8$$
$$33 \times 3$$
$$41 \times 7$$
$$83 \times 4$$

B
$$55 \times 2$$
$$19 \times 5$$
$$97 \times 9$$
$$89 \times 1$$
$$32 \times 1$$

C
$$93 \times 6$$
$$34 \times 8$$
$$45 \times 7$$
$$52 \times 3$$
$$49 \times 5$$

D
$$62 \times 2$$
$$31 \times 4$$
$$75 \times 9$$
$$86 \times 1$$
$$60 \times 5$$

E
$$16 \times 8$$
$$48 \times 7$$
$$96 \times 3$$
$$74 \times 9$$
$$39 \times 2$$

A

$$\begin{array}{r} 47 \\ \times\ 7 \\ \hline \end{array} \qquad \begin{array}{r} 53 \\ \times\ 2 \\ \hline \end{array} \qquad \begin{array}{r} 83 \\ \times\ 6 \\ \hline \end{array} \qquad \begin{array}{r} 77 \\ \times\ 5 \\ \hline \end{array} \qquad \begin{array}{r} 79 \\ \times\ 9 \\ \hline \end{array}$$

B

$$\begin{array}{r} 23 \\ \times\ 1 \\ \hline \end{array} \qquad \begin{array}{r} 22 \\ \times\ 8 \\ \hline \end{array} \qquad \begin{array}{r} 52 \\ \times\ 3 \\ \hline \end{array} \qquad \begin{array}{r} 26 \\ \times\ 4 \\ \hline \end{array} \qquad \begin{array}{r} 12 \\ \times\ 4 \\ \hline \end{array}$$

C

$$\begin{array}{r} 50 \\ \times\ 2 \\ \hline \end{array} \qquad \begin{array}{r} 43 \\ \times\ 6 \\ \hline \end{array} \qquad \begin{array}{r} 17 \\ \times\ 3 \\ \hline \end{array} \qquad \begin{array}{r} 76 \\ \times\ 9 \\ \hline \end{array} \qquad \begin{array}{r} 18 \\ \times\ 8 \\ \hline \end{array}$$

D

$$\begin{array}{r} 94 \\ \times\ 5 \\ \hline \end{array} \qquad \begin{array}{r} 35 \\ \times\ 1 \\ \hline \end{array} \qquad \begin{array}{r} 85 \\ \times\ 7 \\ \hline \end{array} \qquad \begin{array}{r} 57 \\ \times\ 4 \\ \hline \end{array} \qquad \begin{array}{r} 51 \\ \times\ 6 \\ \hline \end{array}$$

E

$$\begin{array}{r} 19 \\ \times\ 2 \\ \hline \end{array} \qquad \begin{array}{r} 45 \\ \times\ 9 \\ \hline \end{array} \qquad \begin{array}{r} 37 \\ \times\ 8 \\ \hline \end{array} \qquad \begin{array}{r} 64 \\ \times\ 1 \\ \hline \end{array} \qquad \begin{array}{r} 62 \\ \times\ 3 \\ \hline \end{array}$$

A

$$\begin{array}{r} 68 \\ \times\ 6 \\ \hline \end{array}$$
$$\begin{array}{r} 62 \\ \times\ 7 \\ \hline \end{array}$$
$$\begin{array}{r} 10 \\ \times\ 8 \\ \hline \end{array}$$
$$\begin{array}{r} 20 \\ \times\ 3 \\ \hline \end{array}$$
$$\begin{array}{r} 88 \\ \times\ 2 \\ \hline \end{array}$$

B

$$\begin{array}{r} 96 \\ \times\ 4 \\ \hline \end{array}$$
$$\begin{array}{r} 86 \\ \times\ 5 \\ \hline \end{array}$$
$$\begin{array}{r} 15 \\ \times\ 9 \\ \hline \end{array}$$
$$\begin{array}{r} 26 \\ \times\ 1 \\ \hline \end{array}$$
$$\begin{array}{r} 46 \\ \times\ 5 \\ \hline \end{array}$$

C

$$\begin{array}{r} 93 \\ \times\ 6 \\ \hline \end{array}$$
$$\begin{array}{r} 23 \\ \times\ 1 \\ \hline \end{array}$$
$$\begin{array}{r} 38 \\ \times\ 9 \\ \hline \end{array}$$
$$\begin{array}{r} 71 \\ \times\ 4 \\ \hline \end{array}$$
$$\begin{array}{r} 51 \\ \times\ 3 \\ \hline \end{array}$$

D

$$\begin{array}{r} 54 \\ \times\ 7 \\ \hline \end{array}$$
$$\begin{array}{r} 21 \\ \times\ 2 \\ \hline \end{array}$$
$$\begin{array}{r} 97 \\ \times\ 8 \\ \hline \end{array}$$
$$\begin{array}{r} 32 \\ \times\ 8 \\ \hline \end{array}$$
$$\begin{array}{r} 56 \\ \times\ 9 \\ \hline \end{array}$$

E

$$\begin{array}{r} 40 \\ \times\ 7 \\ \hline \end{array}$$
$$\begin{array}{r} 65 \\ \times\ 3 \\ \hline \end{array}$$
$$\begin{array}{r} 17 \\ \times\ 1 \\ \hline \end{array}$$
$$\begin{array}{r} 25 \\ \times\ 6 \\ \hline \end{array}$$
$$\begin{array}{r} 64 \\ \times\ 5 \\ \hline \end{array}$$

A
```
    46          17          41          48          73
X    5      X    4      X    3      X    9      X    6
```

B
```
    37          95          25          74          10
X    7      X    1      X    8      X    2      X    5
```

C
```
    99          54          78          64          11
X    6      X    9      X    4      X    1      X    7
```

D
```
    69          87          39          15          65
X    8      X    3      X    2      X    3      X    8
```

E
```
    29          57          77          18          97
X    9      X    7      X    4      X    5      X    1
```

A
$$\begin{array}{r}18\\ \times\ 13\\ \hline\end{array}$$
$$\begin{array}{r}68\\ \times\ 15\\ \hline\end{array}$$
$$\begin{array}{r}74\\ \times\ 10\\ \hline\end{array}$$
$$\begin{array}{r}87\\ \times\ 20\\ \hline\end{array}$$
$$\begin{array}{r}34\\ \times\ 12\\ \hline\end{array}$$

B
$$\begin{array}{r}27\\ \times\ 16\\ \hline\end{array}$$
$$\begin{array}{r}41\\ \times\ 14\\ \hline\end{array}$$
$$\begin{array}{r}46\\ \times\ 17\\ \hline\end{array}$$
$$\begin{array}{r}85\\ \times\ 11\\ \hline\end{array}$$
$$\begin{array}{r}98\\ \times\ 19\\ \hline\end{array}$$

C
$$\begin{array}{r}61\\ \times\ 18\\ \hline\end{array}$$
$$\begin{array}{r}60\\ \times\ 17\\ \hline\end{array}$$
$$\begin{array}{r}22\\ \times\ 15\\ \hline\end{array}$$
$$\begin{array}{r}93\\ \times\ 19\\ \hline\end{array}$$
$$\begin{array}{r}81\\ \times\ 16\\ \hline\end{array}$$

D
$$\begin{array}{r}44\\ \times\ 20\\ \hline\end{array}$$
$$\begin{array}{r}76\\ \times\ 14\\ \hline\end{array}$$
$$\begin{array}{r}24\\ \times\ 18\\ \hline\end{array}$$
$$\begin{array}{r}69\\ \times\ 13\\ \hline\end{array}$$
$$\begin{array}{r}14\\ \times\ 10\\ \hline\end{array}$$

E
$$\begin{array}{r}56\\ \times\ 11\\ \hline\end{array}$$
$$\begin{array}{r}40\\ \times\ 12\\ \hline\end{array}$$
$$\begin{array}{r}30\\ \times\ 20\\ \hline\end{array}$$
$$\begin{array}{r}78\\ \times\ 15\\ \hline\end{array}$$
$$\begin{array}{r}42\\ \times\ 12\\ \hline\end{array}$$

Day: 83
Date:
Score: /25

Name:
Time: :
Rating: ☆☆☆☆☆☆

A

$$\begin{array}{r} 77 \\ \times\ 18 \\ \hline \end{array}$$
$$\begin{array}{r} 46 \\ \times\ 11 \\ \hline \end{array}$$
$$\begin{array}{r} 21 \\ \times\ 19 \\ \hline \end{array}$$
$$\begin{array}{r} 78 \\ \times\ 13 \\ \hline \end{array}$$
$$\begin{array}{r} 73 \\ \times\ 16 \\ \hline \end{array}$$

B

$$\begin{array}{r} 29 \\ \times\ 15 \\ \hline \end{array}$$
$$\begin{array}{r} 88 \\ \times\ 20 \\ \hline \end{array}$$
$$\begin{array}{r} 15 \\ \times\ 14 \\ \hline \end{array}$$
$$\begin{array}{r} 28 \\ \times\ 17 \\ \hline \end{array}$$
$$\begin{array}{r} 95 \\ \times\ 10 \\ \hline \end{array}$$

C

$$\begin{array}{r} 82 \\ \times\ 12 \\ \hline \end{array}$$
$$\begin{array}{r} 35 \\ \times\ 12 \\ \hline \end{array}$$
$$\begin{array}{r} 19 \\ \times\ 16 \\ \hline \end{array}$$
$$\begin{array}{r} 37 \\ \times\ 14 \\ \hline \end{array}$$
$$\begin{array}{r} 20 \\ \times\ 16 \\ \hline \end{array}$$

D

$$\begin{array}{r} 53 \\ \times\ 20 \\ \hline \end{array}$$
$$\begin{array}{r} 44 \\ \times\ 10 \\ \hline \end{array}$$
$$\begin{array}{r} 67 \\ \times\ 18 \\ \hline \end{array}$$
$$\begin{array}{r} 61 \\ \times\ 17 \\ \hline \end{array}$$
$$\begin{array}{r} 13 \\ \times\ 11 \\ \hline \end{array}$$

E

$$\begin{array}{r} 30 \\ \times\ 15 \\ \hline \end{array}$$
$$\begin{array}{r} 62 \\ \times\ 11 \\ \hline \end{array}$$
$$\begin{array}{r} 98 \\ \times\ 11 \\ \hline \end{array}$$
$$\begin{array}{r} 94 \\ \times\ 18 \\ \hline \end{array}$$
$$\begin{array}{r} 89 \\ \times\ 14 \\ \hline \end{array}$$

A
```
   76          42          44          50          86
X  17       X  14       X  13       X  15       X  12
```

B
```
   34          77          81          96          56
X  18       X  11       X  10       X  16       X  20
```

C
```
   23          28          71          19          20
X  19       X  10       X  18       X  15       X  14
```

D
```
   75          87          29          89          14
X  13       X  16       X  11       X  15       X  11
```

E
```
   20          46          52          95          90
X  12       X  15       X  16       X  10       X  20
```

A

| 88 | 86 | 52 | 38 | 59 |
| X 11 | X 17 | X 12 | X 19 | X 13 |

B

| 71 | 40 | 55 | 77 | 82 |
| X 16 | X 14 | X 10 | X 15 | X 18 |

C

| 89 | 66 | 16 | 78 | 72 |
| X 20 | X 13 | X 12 | X 19 | X 12 |

D

| 48 | 32 | 63 | 70 | 21 |
| X 14 | X 20 | X 10 | X 18 | X 11 |

E

| 44 | 85 | 20 | 49 | 36 |
| X 17 | X 15 | X 19 | X 11 | X 13 |

A.
22	33	98	61	50
X 17	X 16	X 18	X 10	X 14

B.
40	39	95	49	78
X 20	X 13	X 11	X 12	X 19

C.
57	60	25	51	16
X 15	X 11	X 15	X 19	X 10

D.
44	76	89	24	67
X 14	X 13	X 20	X 18	X 17

E.
12	73	96	58	99
X 11	X 13	X 10	X 18	X 17

A

59	88	63	35	45
X 14	X 20	X 16	X 10	X 18

B

49	19	21	26	54
X 13	X 18	X 12	X 15	X 17

C

92	58	84	94	85
X 11	X 14	X 19	X 18	X 15

D

17	97	33	16	90
X 10	X 12	X 13	X 14	X 11

E

91	73	68	20	57
X 20	X 15	X 11	X 16	X 19

A
```
     93          85          46          43          92
  x  20       x  16       x  17       x  13       x  12
```

B
```
     98          24          47          99          70
  x  11       x  10       x  18       x  14       x  15
```

C
```
     69          81          15          16          74
  x  19       x  20       x  11       x  12       x  13
```

D
```
     39          72          83          77          41
  x  14       x  12       x  10       x  18       x  19
```

E
```
     33          62          25          96          64
  x  15       x  14       x  11       x  17       x  19
```

A

72	40	86	60	74
X 14	X 11	X 13	X 10	X 16

B

21	73	80	20	55
X 17	X 19	X 18	X 17	X 12

C

16	52	92	47	96
X 15	X 11	X 16	X 18	X 20

D

49	26	87	20	41
X 19	X 12	X 17	X 13	X 10

E

70	48	44	50	14
X 15	X 11	X 20	X 18	X 10

A

80	23	35	58	79
X 20	X 16	X 11	X 10	X 19

B

88	51	18	15	89
X 12	X 17	X 14	X 12	X 13

C

49	91	42	24	86
X 18	X 16	X 20	X 19	X 11

D

92	38	31	50	59
X 18	X 17	X 14	X 15	X 12

E

60	72	43	64	99
X 10	X 15	X 13	X 20	X 12

A
```
    80          91          86          98          90
X   12      X   14      X   20      X   13      X   17
```

B
```
    64          94          44          70          29
X   18      X   15      X   11      X   19      X   10
```

C
```
    20          32          69          38          81
X   16      X   18      X   16      X   12      X   11
```

D
```
    36          47          45          18          56
X   19      X   17      X   20      X   13      X   10
```

E
```
    75          96          42          26          65
X   14      X   11      X   19      X   18      X   16
```

A

90	74	64	44	98
X 33	X 39	X 55	X 14	X 86

B

78	81	75	39	69
X 17	X 66	X 65	X 22	X 19

C

80	48	45	54	33
X 52	X 45	X 43	X 28	X 12

D

72	95	94	93	83
X 51	X 26	X 72	X 89	X 46

E

90	97	67	67	92
X 38	X 77	X 10	X 16	X 60

A

85	29	38	91	50
X 30	X 24	X 34	X 42	X 21

B

73	57	92	48	37
X 16	X 40	X 73	X 21	X 25

C

59	97	55	92	46
X 36	X 59	X 19	X 11	X 41

D

53	52	69	96	76
X 23	X 22	X 51	X 58	X 61

E

87	75	38	43	79
X 36	X 41	X 23	X 26	X 21

Day: 94

Date:

Score: /25

Name:

Time: :

Rating: ☆☆☆☆☆☆

A
| 48 | 88 | 85 | 34 | 35 |
| X 43 | X 83 | X 56 | X 18 | X 28 |

B
| 99 | 50 | 35 | 40 | 45 |
| X 86 | X 12 | X 25 | X 27 | X 37 |

C
| 49 | 36 | 82 | 82 | 19 |
| X 44 | X 17 | X 66 | X 20 | X 15 |

D
| 75 | 65 | 31 | 98 | 91 |
| X 32 | X 40 | X 31 | X 90 | X 52 |

E
| 56 | 68 | 91 | 84 | 63 |
| X 42 | X 57 | X 60 | X 10 | X 13 |

A

76	75	70	30	47
X 16	X 71	X 25	X 13	X 15

B

58	78	81	93	96
X 22	X 73	X 48	X 78	X 64

C

74	63	72	62	46
X 49	X 47	X 29	X 42	X 27

D

29	90	58	86	39
X 25	X 43	X 22	X 61	X 19

E

95	94	56	37	82
X 39	X 68	X 32	X 37	X 80

A

79	89	68	41	85
X 33	X 15	X 49	X 28	X 35

B

79	76	88	70	59
X 48	X 60	X 18	X 28	X 24

C

85	79	71	83	60
X 13	X 55	X 63	X 18	X 50

D

92	55	81	76	89
X 46	X 20	X 48	X 54	X 82

E

88	63	41	70	36
X 11	X 26	X 37	X 21	X 19

A
```
    18          53          85          77          62
X   17      X   26      X   21      X   74      X   44
```

B
```
    75          62          30          58          97
X   66      X   12      X   12      X   23      X   42
```

C
```
    31          54          86          34          90
X   30      X   27      X   47      X   10      X   52
```

D
```
    94          89          51          91          69
X   57      X   72      X   13      X   16      X   64
```

E
```
    93          92          41          84          52
X   32      X   53      X   40      X   26      X   24
```

A

90	67	23	77	98
X 73	X 61	X 19	X 40	X 71

B

50	72	91	84	31
X 47	X 10	X 55	X 38	X 29

C

75	17	89	82	76
X 34	X 10	X 45	X 16	X 49

D

69	81	52	42	87
X 56	X 60	X 46	X 37	X 65

E

74	96	80	64	35
X 33	X 25	X 78	X 58	X 14

A

99	98	41	32	87
X 82	X 96	X 19	X 29	X 58

B

61	47	44	86	91
X 32	X 27	X 34	X 52	X 66

C

84	82	84	95	87
X 41	X 57	X 70	X 89	X 11

D

71	50	46	75	31
X 26	X 29	X 11	X 19	X 15

E

60	99	39	85	85
X 56	X 68	X 20	X 53	X 18

A

64	69	59	65	67
X 28	X 52	X 40	X 43	X 63

B

98	95	77	64	61
X 78	X 33	X 61	X 62	X 37

C

53	77	90	79	94
X 36	X 38	X 80	X 69	X 93

D

36	70	75	97	51
X 25	X 40	X 13	X 44	X 45

E

26	56	98	30	82
X 18	X 15	X 21	X 10	X 11

A

75	87	89	24	73
X 43	X 43	X 18	X 19	X 41

B

42	60	81	35	57
X 32	X 55	X 42	X 33	X 30

C

60	70	52	97	99
X 40	X 54	X 31	X 94	X 87

D

65	69	72	88	91
X 27	X 14	X 28	X 82	X 46

E

97	48	66	79	60
X 18	X 24	X 23	X 59	X 33

Multiplication Answer Key Sheet (1/2)

DAY	Row					
1	A	0,	6,	9,	0,	0
	B	8,	0,	7,	0,	0
	C	0,	0,	1,	2,	0
	D	5,	0,	0,	0,	0
	E	4,	3,	6,	7,	0
2	A	15,	6,	3,	0,	10
	B	10,	0,	27,	16,	8
	C	24,	14,	21,	12,	12
	D	9,	4,	0,	18,	6
	E	18,	24,	12,	8,	2
3	A	25,	5,	15,	10,	4
	B	15,	24,	35,	0,	20
	C	20,	8,	0,	20,	12
	D	40,	36,	28,	16,	24
	E	30,	45,	28,	10,	32
4	A	49,	42,	14,	0,	18
	B	35,	48,	21,	54,	42
	C	18,	30,	28,	63,	12
	D	35,	36,	24,	0,	6
	E	7,	56,	42,	28,	48
5	A	81,	63,	0,	27,	40
	B	36,	0,	18,	72,	0
	C	8,	8,	9,	36,	56
	D	54,	64,	48,	24,	16
	E	45,	72,	64,	54,	32
6	A	2,	5,	3,	8,	0
	B	5,	0,	9,	4,	0
	C	0,	0,	1,	3,	0
	D	7,	0,	0,	0,	0
	E	7,	7,	6,	1,	0
7	A	24,	3,	6,	15,	8
	B	10,	18,	9,	8,	16
	C	21,	2,	27,	27,	6
	D	24,	18,	10,	12,	4
	E	21,	18,	8,	6,	14
8	A	30,	15,	20,	45,	36
	B	40,	20,	40,	8,	4
	C	12,	10,	5,	20,	8
	D	35,	32,	24,	12,	28
	E	35,	25,	36,	30,	20
9	A	56,	14,	49,	42,	6
	B	56,	30,	63,	6,	30
	C	6,	48,	21,	28,	24
	D	42,	36,	42,	12,	18
	E	35,	28,	54,	49,	54
10	A	81,	72,	27,	36,	56
	B	9,	72,	45,	8,	72
	C	16,	8,	63,	63,	16
	D	72,	48,	64,	24,	32
	E	18,	54,	32,	45,	40
11	A	12,	4,	16,	10,	5
	B	12,	9,	8,	9,	10
	C	3,	1,	20,	14,	8
	D	2,	3,	4,	2,	6
	E	18,	6,	4,	4,	7
12	A	32,	4,	36,	8,	24
	B	12,	27,	12,	18,	9
	C	24,	30,	16,	20,	3
	D	28,	27,	15,	12,	21
	E	24,	40,	21,	16,	6
13	A	24,	42,	18,	48,	5
	B	6,	10,	36,	40,	50
	C	20,	30,	30,	60,	10
	D	12,	35,	45,	15,	25
	E	54,	6,	40,	36,	20
14	A	80,	24,	8,	32,	40
	B	72,	14,	72,	21,	7
	C	49,	35,	16,	48,	28
	D	56,	56,	63,	42,	70
	E	40,	64,	28,	40,	14
15	A	60,	40,	80,	30,	27
	B	80,	36,	10,	45,	90
	C	18,	54,	100,	50,	81
	D	70,	18,	36,	36,	72
	E	90,	20,	45,	60,	63
16	A	10,	0,	0,	0,	6
	B	4,	1,	0,	0,	5
	C	0,	0,	2,	5,	0
	D	1,	0,	2,	3,	0
	E	0,	0,	0,	9,	10
17	A	8,	18,	4,	20,	6
	B	14,	6,	10,	20,	2
	C	10,	4,	16,	14,	9
	D	4,	5,	12,	3,	2
	E	2,	8,	9,	7,	20
18	A	4,	5,	2,	9,	6
	B	24,	2,	30,	30,	18
	C	5,	9,	0,	12,	8
	D	18,	21,	6,	1,	9
	E	6,	9,	6,	6,	20
19	A	9,	18,	8,	4,	6
	B	40,	6,	4,	2,	24
	C	10,	32,	4,	36,	5
	D	9,	6,	14,	18,	8
	E	4,	27,	32,	18,	8
20	A	24,	4,	7,	20,	20
	B	5,	9,	30,	9,	4
	C	10,	20,	15,	50,	8
	D	24,	7,	12,	27,	4
	E	6,	9,	40,	24,	10
21	A	35,	30,	8,	15,	16
	B	12,	10,	5,	8,	45
	C	9,	10,	1,	24,	2
	D	50,	12,	42,	8,	2
	E	4,	50,	45,	2,	32

DAY	Row					
22	A	56,	3,	10,	6,	35
	B	25,	9,	36,	18,	8
	C	60,	35,	20,	14,	32
	D	3,	6,	63,	12,	6
	E	20,	7,	30,	45,	8
23	A	70,	14,	16,	24,	45
	B	4,	36,	42,	25,	48
	C	12,	3,	27,	20,	20
	D	8,	56,	8,	21,	15
	E	32,	9,	6,	24,	6
24	A	63,	25,	12,	36,	60
	B	64,	18,	3,	12,	8
	C	50,	12,	7,	63,	40
	D	8,	30,	24,	14,	16
	E	6,	14,	40,	42,	28
25	A	28,	35,	80,	15,	36
	B	9,	18,	20,	18,	6
	C	18,	5,	12,	6,	60
	D	42,	56,	40,	10,	12
	E	12,	70,	4,	40,	24
26	A	56,	5,	50,	8,	14
	B	18,	36,	72,	18,	10
	C	40,	14,	40,	24,	90
	D	14,	63,	12,	6,	9
	E	24,	50,	6,	100,	40
27	A	18,	20,	40,	4,	7
	B	15,	18,	63,	48,	16
	C	28,	6,	80,	10,	1
	D	10,	24,	24,	81,	42
	E	64,	20,	14,	45,	3
28	A	8,	32,	2,	6,	35
	B	48,	50,	81,	9,	70
	C	56,	27,	6,	24,	40
	D	8,	56,	45,	6,	5
	E	7,	14,	24,	3,	30
29	A	9,	45,	36,	24,	63
	B	10,	28,	36,	8,	10
	C	25,	24,	4,	28,	7
	D	32,	81,	12,	3,	60
	E	16,	54,	27,	50,	6
30	A	36,	9,	10,	9,	56
	B	36,	14,	48,	50,	2
	C	5,	36,	16,	49,	5
	D	63,	24,	24,	24,	40
	E	60,	15,	21,	4,	54
31	A	30,	5,	32,	70,	16
	B	54,	27,	12,	2,	70
	C	6,	8,	45,	14,	25
	D	32,	72,	10,	42,	9
	E	80,	49,	9,	20,	6
32	A	54,	4,	7,	30,	9
	B	18,	24,	64,	35,	15
	C	48,	70,	6,	27,	8
	D	56,	42,	25,	12,	8
	E	6,	5,	7,	36,	54
33	A	16,	6,	45,	28,	6
	B	6,	50,	24,	72,	15
	C	10,	28,	30,	18,	27
	D	4,	42,	24,	40,	4
	E	15,	36,	64,	90,	16
34	A	21,	10,	15,	16,	12
	B	56,	9,	54,	2,	80
	C	28,	16,	35,	1,	81
	D	12,	12,	64,	20,	10
	E	72,	35,	27,	35,	100
35	A	9,	36,	21,	30,	5
	B	80,	6,	42,	12,	16
	C	36,	32,	16,	6,	35
	D	4,	63,	30,	10,	4
	E	48,	30,	45,	9,	4
36	A	12,	8,	1,	10,	4
	B	6,	7,	11,	4,	20
	C	8,	5,	20,	3,	24
	D	12,	2,	22,	10,	14
	E	6,	18,	16,	2,	9
37	A	8,	27,	36,	12,	33
	B	40,	24,	15,	24,	24
	C	48,	9,	44,	21,	12
	D	6,	28,	32,	16,	20
	E	30,	36,	4,	3,	18
38	A	36,	35,	40,	30,	20
	B	18,	50,	10,	42,	60
	C	24,	15,	48,	60,	12
	D	5,	30,	72,	66,	54
	E	55,	6,	60,	25,	45
39	A	56,	49,	7,	42,	77
	B	48,	35,	14,	24,	56
	C	16,	63,	96,	84,	8
	D	56,	32,	72,	40,	64
	E	28,	80,	88,	21,	70
40	A	50,	90,	81,	99,	54
	B	30,	18,	27,	60,	100
	C	100,	108,	70,	36,	120
	D	63,	10,	96,	121,	90
	E	9,	30,	40,	72,	45
41	A	84,	77,	55,	22,	66
	B	120,	44,	121,	24,	120
	C	132,	11,	12,	33,	60
	D	99,	108,	96,	36,	48
	E	132,	72,	96,	88,	88
42	A	64,	63,	12,	9,	11
	B	54,	30,	30,	4,	24
	C	14,	100,	18,	100,	49
	D	44,	9,	24,	20,	36
	E	48,	5,	32,	20,	30

DAY	Row					
43	A	10,	12,	15,	8,	45
	B	42,	2,	48,	44,	5
	C	72,	28,	16,	42,	30
	D	24,	54,	72,	9,	55
	E	2,	21,	12,	80,	45
44	A	12,	45,	4,	21,	30
	B	33,	36,	8,	80,	6
	C	84,	55,	10,	48,	3
	D	8,	63,	16,	63,	18
	E	24,	110,	48,	35,	45
45	A	5,	20,	28,	16,	30
	B	27,	40,	4,	6,	21
	C	84,	99,	18,	24,	7
	D	45,	96,	66,	32,	18
	E	6,	30,	55,	1,	90
46	A	12,	28,	14,	1,	99
	B	132,	30,	48,	25,	60
	C	60,	16,	10,	10,	63
	D	28,	22,	8,	56,	3
	E	81,	18,	48,	15,	15
47	A	22,	4,	84,	20,	50
	B	3,	42,	48,	64,	33
	C	45,	50,	21,	4,	9
	D	12,	35,	16,	6,	32
	E	72,	80,	54,	6,	45
48	A	36,	44,	24,	45,	3
	B	42,	7,	56,	60,	12
	C	4,	20,	32,	27,	77
	D	4,	10,	12,	70,	100
	E	24,	21,	40,	24,	8
49	A	16,	28,	18,	15,	54
	B	20,	20,	27,	108,	48
	C	70,	1,	81,	24,	9
	D	77,	35,	16,	42,	32
	E	80,	110,	11,	35,	28
50	A	120,	99,	12,	12,	27
	B	10,	35,	16,	36,	63
	C	56,	44,	10,	2,	32
	D	48,	6,	16,	70,	20
	E	55,	72,	27,	110,	35
51	A	72,	30,	42,	9,	4
	B	30,	50,	33,	4,	6
	C	120,	72,	24,	12,	49
	D	66,	4,	24,	50,	120
	E	108,	36,	10,	11,	40
52	A	66,	180,	19,	32,	14
	B	91,	48,	90,	30,	0
	C	153,	30,	153,	144,	22
	D	95,	52,	20,	0,	140
	E	40,	96,	10,	45,	12
53	A	0,	120,	112,	26,	152
	B	153,	72,	50,	42,	34
	C	33,	0,	16,	120,	99
	D	108,	65,	56,	140,	55
	E	152,	126,	64,	0,	36
54	A	0,	105,	99,	19,	102
	B	90,	32,	160,	56,	30
	C	80,	135,	14,	0,	95
	D	36,	120,	52,	112,	140
	E	99,	56,	114,	20,	15
55	A	90,	0,	112,	56,	55
	B	144,	114,	36,	34,	114
	C	52,	126,	20,	90,	128
	D	52,	0,	51,	55,	119
	E	24,	0,	96,	30,	60
56	A	48,	180,	52,	0,	66
	B	17,	30,	112,	126,	51
	C	80,	72,	0,	140,	70
	D	40,	32,	117,	15,	75
	E	44,	98,	40,	38,	144
57	A	112,	30,	52,	51,	95
	B	0,	126,	120,	11,	72
	C	30,	40,	60,	114,	153
	D	15,	128,	70,	0,	48
	E	13,	108,	84,	126,	84
58	A	100,	144,	39,	133,	144
	B	34,	60,	10,	0,	108
	C	0,	119,	64,	90,	60
	D	33,	152,	26,	15,	108
	E	80,	0,	105,	144,	16
59	A	140,	90,	153,	144,	39
	B	55,	0,	24,	40,	33
	C	0,	96,	95,	16,	102
	D	126,	20,	126,	52,	36
	E	75,	39,	90,	99,	90
60	A	64,	45,	100,	114,	14
	B	88,	91,	36,	0,	20
	C	85,	135,	33,	0,	80
	D	48,	98,	102,	38,	77
	E	48,	70,	17,	60,	60
61	A	144,	95,	117,	0,	20
	B	84,	33,	84,	34,	70
	C	96,	152,	0,	117,	70
	D	30,	64,	60,	11,	96
	E	50,	126,	68,	132,	
62	A	837,	140,	164,	384,	132
	B	518,	288,	76,	60,	168
	C	108,	612,	266,	90,	63
	D	396,	664,	165,	686,	69
	E	86,	385,	594,	12,	536
63	A	462,	410,	344,	92,	124
	B	48,	540,	408,	96,	74
	C	355,	117,	64,	232,	534
	D	196,	456,	114,	20,	72
	E	208,	92,	252,	392,	243

DAY						
64	A	46,	425,	84,	276,	81
	B	584,	240,	819,	644,	415
	C	201,	416,	70,	198,	80
	D	336,	136,	531,	210,	117
	E	408,	45,	105,	105,	204
65	A	252,	776,	213,	182,	348
	B	108,	50,	86,	120,	712
	C	208,	19,	558,	189,	162
	D	295,	168,	525,	368,	450
	E	93,	60,	72,	328,	207
66	A	360,	576,	657,	150,	72
	B	126,	324,	120,	71,	378
	C	138,	204,	35,	80,	882
	D	166,	168,	420,	98,	594
	E	290,	90,	696,	53,	312
67	A	82,	175,	312,	30,	192
	B	217,	765,	236,	480,	222
	C	252,	405,	696,	325,	138
	D	280,	159,	44,	86,	58
	E	581,	170,	228,	420,	129
68	A	62,	128,	270,	77,	765
	B	198,	448,	235,	88,	282
	C	98,	174,	720,	576,	54
	D	332,	161,	305,	57,	81
	E	477,	408,	42,	256,	295
69	A	325,	129,	480,	120,	60
	B	98,	264,	621,	287,	249
	C	435,	630,	198,	352,	315
	D	344,	576,	73,	280,	171
	E	294,	275,	24,	54,	280
70	A	76,	336,	65,	576,	582
	B	329,	138,	240,	140,	63
	C	567,	189,	154,	258,	340
	D	456,	296,	120,	36,	415
	E	504,	474,	462,	128,	26
71	A	204,	94,	776,	308,	185
	B	153,	711,	128,	637,	248
	C	159,	180,	712,	380,	774
	D	270,	72,	304,	639,	301
	E	396,	297,	490,	300,	60
72	A	192,	136,	325,	855,	534
	B	240,	74,	231,	476,	275
	C	657,	36,	264,	200,	244
	D	376,	392,	210,	182,	60
	E	288,	522,	336,	192,	22
73	A	440,	237,	204,	52,	120
	B	47,	760,	126,	546,	204
	C	100,	406,	144,	744,	333
	D	220,	132,	27,	581,	243
	E	240,	75,	11,	182,	388
74	A	40,	200,	69,	553,	387
	B	534,	416,	285,	68,	497
	C	648,	54,	130,	86,	738
	D	504,	244,	400,	63,	28
	E	486,	372,	116,	120,	325
75	A	282,	216,	168,	756,	95
	B	126,	316,	33,	73,	504
	C	88,	348,	125,	230,	468
	D	570,	156,	128,	693,	287
	E	324,	150,	50,	276,	65
76	A	87,	205,	480,	74,	36
	B	175,	774,	256,	184,	132
	C	292,	891,	160,	250,	182
	D	490,	117,	10,	47,	120
	E	364,	95,	276,	450,	33
77	A	273,	190,	388,	88,	21
	B	90,	828,	184,	217,	84
	C	350,	248,	256,	792,	166
	D	162,	35,	215,	145,	238
	E	98,	144,	188,	174,	284
78	A	462,	96,	99,	287,	332
	B	110,	95,	873,	89,	32
	C	558,	272,	315,	156,	245
	D	124,	124,	675,	86,	300
	E	128,	336,	288,	666,	78
79	A	329,	106,	498,	385,	711
	B	23,	176,	156,	104,	48
	C	100,	258,	51,	684,	144
	D	470,	35,	595,	228,	306
	E	38,	405,	296,	64,	186
80	A	408,	434,	80,	60,	176
	B	384,	430,	135,	26,	230
	C	558,	23,	342,	284,	153
	D	378,	42,	776,	256,	504
	E	280,	195,	17,	150,	320
81	A	230,	68,	123,	432,	438
	B	259,	95,	200,	148,	50
	C	594,	486,	312,	64,	77
	D	552,	261,	78,	45,	520
	E	261,	399,	308,	90,	97
82	A	234,	1020,	740,	1740,	408
	B	432,	574,	782,	935,	1862
	C	1098,	1020,	330,	1767,	1296
	D	880,	1064,	432,	897,	140
	E	616,	480,	600,	1170,	504
83	A	1386,	506,	399,	1014,	1168
	B	435,	1760,	210,	476,	950
	C	984,	420,	304,	518,	320
	D	1060,	440,	1206,	1037,	143
	E	450,	682,	1078,	1692,	1246
84	A	1292,	588,	572,	750,	1032
	B	612,	847,	810,	1536,	1120
	C	437,	280,	1278,	285,	280
	D	975,	1392,	319,	1335,	154
	E	240,	690,	832,	950,	1800

DAY						
85	A	968,	1462,	624,	722,	767
	B	1136,	560,	550,	1155,	1476
	C	1780,	858,	192,	1482,	864
	D	672,	640,	630,	1260,	231
	E	748,	1275,	380,	539,	468
86	A	374,	528,	1764,	610,	700
	B	800,	507,	1045,	588,	1482
	C	855,	660,	375,	969,	160
	D	616,	988,	1780,	432,	1139
	E	132,	949,	960,	1044,	1683
87	A	826,	1760,	1008,	350,	810
	B	637,	342,	252,	390,	918
	C	1012,	812,	1596,	1692,	1275
	D	170,	1164,	429,	224,	990
	E	1820,	1095,	748,	320,	1083
88	A	1860,	1360,	782,	559,	1104
	B	1078,	240,	846,	1386,	1050
	C	1311,	1620,	165,	192,	962
	D	546,	864,	830,	1386,	779
	E	495,	868,	275,	1632,	1216
89	A	1008,	440,	1118,	600,	1184
	B	357,	1387,	1440,	340,	660
	C	240,	572,	1472,	846,	1920
	D	931,	312,	1479,	260,	410
	E	1050,	528,	880,	900,	140
90	A	1600,	368,	385,	580,	1501
	B	1056,	867,	252,	180,	1157
	C	882,	1456,	840,	456,	946
	D	1656,	646,	434,	750,	708
	E	600,	1080,	559,	1280,	1188
91	A	960,	1274,	1720,	1274,	1530
	B	1152,	1410,	484,	1330,	290
	C	320,	576,	1104,	456,	891
	D	684,	799,	900,	234,	560
	E	1050,	1056,	798,	468,	1040
92	A	2970,	2886,	3520,	616,	8428
	B	1326,	5346,	4875,	858,	1311
	C	4160,	2160,	1935,	1512,	396
	D	3672,	2470,	6768,	8277,	3818
	E	3420,	7469,	670,	1072,	5520
93	A	2550,	696,	1292,	3822,	1050
	B	1168,	2280,	6716,	1008,	925
	C	2124,	5723,	1045,	1012,	1886
	D	1219,	1144,	3519,	5568,	4636
	E	3132,	3075,	874,	1118,	1659
94	A	2064,	7304,	4760,	612,	980
	B	8514,	600,	875,	1080,	1665
	C	2156,	612,	5412,	1640,	285
	D	2400,	2600,	961,	8820,	4732
	E	2352,	3876,	5460,	840,	819
95	A	1216,	5325,	1750,	390,	705
	B	1276,	5694,	3888,	7254,	6144
	C	3626,	2961,	2088,	2604,	1242
	D	725,	3870,	1276,	5246,	741
	E	3705,	6392,	1792,	1369,	6560
96	A	2607,	1335,	3332,	1148,	2975
	B	3792,	4560,	1584,	1960,	1416
	C	1105,	4345,	4473,	1494,	3000
	D	4232,	1100,	3888,	4104,	7298
	E	968,	1638,	1517,	1470,	684
97	A	306,	1378,	1785,	5698,	2728
	B	4950,	744,	360,	1334,	4074
	C	930,	1458,	4042,	340,	4680
	D	5358,	6408,	663,	1456,	4416
	E	2976,	4876,	1640,	2184,	1248
98	A	6570,	4087,	437,	3080,	6958
	B	2350,	720,	5005,	3192,	899
	C	2550,	170,	4005,	1312,	3724
	D	3864,	4860,	2392,	1554,	5655
	E	2442,	2400,	6240,	3712,	490
99	A	8118,	9408,	779,	928,	5046
	B	1952,	1269,	1496,	4472,	6006
	C	3444,	4674,	5880,	8455,	957
	D	1846,	1450,	506,	1425,	465
	E	3360,	6732,	780,	4505,	1530
100	A	1792,	3588,	2360,	2795,	4221
	B	7644,	3135,	4697,	3968,	2257
	C	1908,	2926,	7200,	5451,	8742
	D	900,	2800,	975,	4268,	2295
	E	468,	840,	2058,	300,	902
101	A	3225,	3741,	1602,	456,	2993
	B	1344,	3300,	3402,	1155,	1710
	C	2400,	3780,	1612,	9118,	8613
	D	1755,	966,	2016,	7216,	4186
	E	1746,	1152,	1518,	4661,	1980

Answer Key
Mapping:

DAY 1

Day 1
Problems
& Solutions

DAY						
1	A	0,	6,	9,	0,	0
	B	8,	0,	7,	2,	0
	C	0,	0,	1,	0,	0
	D	5,	0,	0,	0,	0
	E	4,	3,	6,	7,	0

Five rows (A-E) with five columns of solutions for each day.

$$\begin{array}{r} A \\ \times\ 0 \\ \hline 0 \end{array} \qquad \begin{array}{r} 8 \\ \times\ 0 \\ \hline 0 \end{array} \qquad \begin{array}{r} 6 \\ \times\ 1 \\ \hline 6 \end{array} \qquad \begin{array}{r} 9 \\ \times\ 1 \\ \hline 9 \end{array} \qquad \begin{array}{r} 0 \\ \times\ 1 \\ \hline 0 \end{array} \qquad \begin{array}{r} 7 \\ \times\ 0 \\ \hline 0 \end{array}$$

Multiplication Tables Sheet

1 Times Table
1 x 1 = 1
1 x 2 = 2
1 x 3 = 3
1 x 4 = 4
1 x 5 = 5
1 x 6 = 6
1 x 7 = 7
1 x 8 = 8
1 x 9 = 9
1 x 10 = 10
1 x 11 = 11
1 x 12 = 12

2 Times Table
2 x 1 = 2
2 x 2 = 4
2 x 3 = 6
2 x 4 = 8
2 x 5 = 10
2 x 6 = 12
2 x 7 = 14
2 x 8 = 16
2 x 9 = 18
2 x 10 = 20
2 x 11 = 22
2 x 12 = 24

3 Times Table
3 x 1 = 3
3 x 2 = 6
3 x 3 = 9
3 x 4 = 12
3 x 5 = 15
3 x 6 = 18
3 x 7 = 21
3 x 8 = 24
3 x 9 = 27
3 x 10 = 30
3 x 11 = 33
3 x 12 = 36

4 Times Table
4 x 1 = 4
4 x 2 = 8
4 x 3 = 12
4 x 4 = 16
4 x 5 = 20
4 x 6 = 24
4 x 7 = 28
4 x 8 = 32
4 x 9 = 36
4 x 10 = 40
4 x 11 = 44
4 x 12 = 48

5 Times Table
5 x 1 = 5
5 x 2 = 10
5 x 3 = 15
5 x 4 = 20
5 x 5 = 25
5 x 6 = 30
5 x 7 = 35
5 x 8 = 40
5 x 9 = 45
5 x 10 = 50
5 x 11 = 55
5 x 12 = 60

6 Times Table
6 x 1 = 6
6 x 2 = 12
6 x 3 = 18
6 x 4 = 24
6 x 5 = 30
6 x 6 = 36
6 x 7 = 42
6 x 8 = 48
6 x 9 = 54
6 x 10 = 60
6 x 11 = 66
6 x 12 = 72

7 Times Table
7 x 1 = 7
7 x 2 = 14
7 x 3 = 21
7 x 4 = 28
7 x 5 = 35
7 x 6 = 42
7 x 7 = 49
7 x 8 = 56
7 x 9 = 63
7 x 10 = 70
7 x 11 = 77
7 x 12 = 84

8 Times Table
8 x 1 = 8
8 x 2 = 16
8 x 3 = 24
8 x 4 = 32
8 x 5 = 40
8 x 6 = 48
8 x 7 = 56
8 x 8 = 64
8 x 9 = 72
8 x 10 = 80
8 x 11 = 88
8 x 12 = 96

9 Times Table
9 x 1 = 9
9 x 2 = 18
9 x 3 = 27
9 x 4 = 36
9 x 5 = 45
9 x 6 = 54
9 x 7 = 63
9 x 8 = 72
9 x 9 = 81
9 x 10 = 90
9 x 11 = 99
9 x 12 = 108

10 Times Table
10 x 1 = 10
10 x 2 = 20
10 x 3 = 30
10 x 4 = 40
10 x 5 = 50
10 x 6 = 60
10 x 7 = 70
10 x 8 = 80
10 x 9 = 90
10 x 10 = 100
10 x 11 = 110
10 x 12 = 120

11 Times Table
11 x 1 = 11
11 x 2 = 22
11 x 3 = 33
11 x 4 = 44
11 x 5 = 55
11 x 6 = 66
11 x 7 = 77
11 x 8 = 88
11 x 9 = 99
11 x 10 = 110
11 x 11 = 121
11 x 12 = 132

12 Times Table
12 x 1 = 12
12 x 2 = 24
12 x 3 = 36
12 x 4 = 48
12 x 5 = 60
12 x 6 = 72
12 x 7 = 84
12 x 8 = 96
12 x 9 = 108
12 x 10 = 120
12 x 11 = 132
12 x 12 = 144

6000 Practice Problems

100 Days of Timed Tests

HORIZONTAL MULTIPLICATION & DIVISION

Facts 1-12

Ages 7-10

109 Pages

Grade 3-5

MATH DRILLS   SPEED MATH PRACTICE

ISBN: 9798887200125

6600 Practice Problems

100+ Days of Timed Tests

Double Digit Addition & Subtraction

Grade 1-3

60

120 Pages

Ages 6-9

MATH DRILLS WITH & WITHOUT REGROUPING

ISBN: 9798887200200

3000+ Practice Problems

100+ Days of Timed Tests

Triple Digit Addition & Subtraction

Math Excel

Ages 7-9

110 Pages

Grade 2-3

MATH DRILLS WITH & WITHOUT REGROUPING

ISBN: 9798887200217

High Frequency Words Tracing

100 MUST KNOW SIGHT WORDS

PRACTICE WORKBOOK

Strokes Uppercase Lowercase

Tracing Coloring Sentences

For Kids: Pre-K, K & Grade 1

Aa Aa

can

I can do it.

Ages 3+

WRITING   COLORING

ISBN: 9798887200132

Certificate of Excellence Award

Single & Double Digit Multiplication
(With Regrouping)

Congratulations!

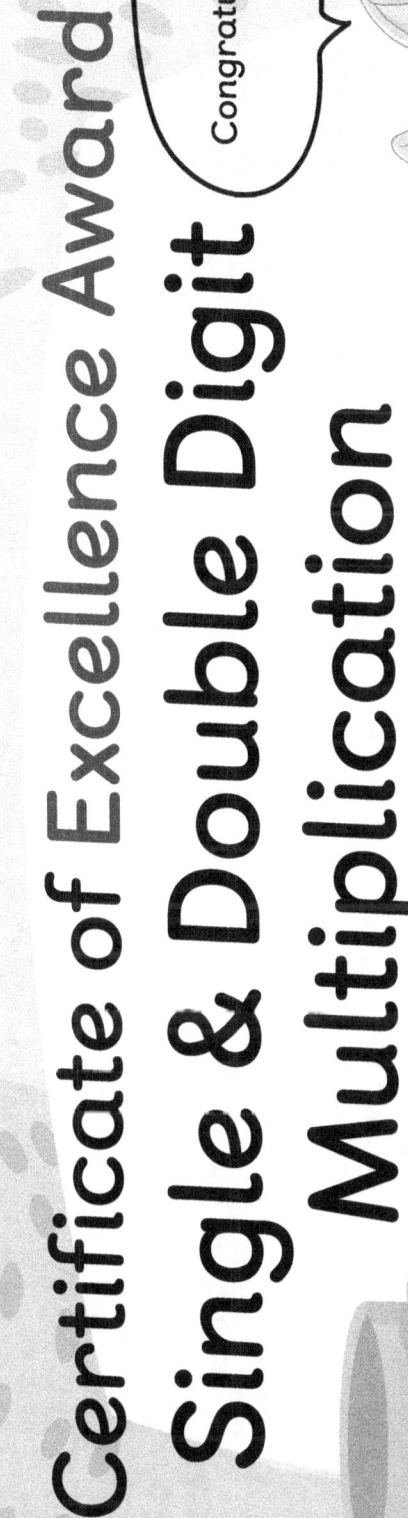

You are a SuperStar!

By:

abcZbook

Date: